I0467820

November 2015

www.UltimateBrainstorming.com

ISBN-13:
978-1519495112

ISBN-10:
1519495110

This book, designed for boys and girls, is published with my consent.

Thomas A Edison

EDISON AT WORK IN ONE OF THE CHEMICAL
ROOMS AT THE ORANGE LABORATORY

THE BOYS' LIFE OF EDISON

BY

WILLIAM H. MEADOWCROFT

OF THE EDISON LABORATORY. AUTHOR OF "ABC OF ELECTRICITY," "ABC OF THE X-RAYS"

WITH AUTOBIOGRAPHICAL NOTES

BY MR. EDISON

ILLUSTRATED

HARPER & BROTHERS PUBLISHERS

NEW YORK AND LONDON

CONTENTS

CHAP.

ILLUSTRATIONS

INTRODUCTION

This is the story of a great inventor, the most conspicuous figure of the age of electricity.

The story is largely autobiography, for, through the author's association with Mr. Edison, it has been possible often to obtain his own narrative of his life. For nearly thirty-one years the author has had the privilege of a connection with Mr. Edison and the Edison companies, and at present he is acting as Mr. Edison's assistant. Every page of the book has been read by Mr. Edison himself, and it is published with his approval as the authoritative story of his life to the present time.

It is probably as a worker of wonders, an interpreter of the secrets of Nature, an actual wizard of science, that Edison fascinates the imagination of almost every boy. In this picture of the actual facts of the inventor's life the reader will find that while Edison is just as great as imagined, yet this greatness has not been reached by chance, but honestly earned by the hardest kind of hard work and the most intense and earnest application. The wonderful things that he has accomplished have been the things that he purposely set out to do, and are not the result of some happy thought, or blind luck, or chance.

There has been but little abatement in Mr. Edison's activities. The flight of time has not dimmed his vivid imagination; has brought no change to his clear broad mental vision; nor has his capacity for intensive, forceful work perceptibly lessened. There is no telling what other inventions he may yet make to benefit the world, but if he never added anything to what he has already done, his life and achievements afford the telling of one of the most remarkable stories in the history of the world.

The author has had the honor and pleasure of assisting in the preparation of a large and comprehensive biography entitled, *Edison: His Life and Inventions*, by Frank L. Dyer and T. Commerford Martin, published by the publishers of the present volume. He gratefully acknowledges the fact that certain features of this book have been adapted from the pages of that elaborate biography. For the permission to do this he tenders his thanks to his friends Frank L. Dyer and the late T. Commerford Martin.

WILLIAM H. MEADOWCROFT.

THE BOYS' LIFE OF EDISON

I

THE EARLY DAYS OF ELECTRICITY

This is the life story of the greatest of inventors in the field of electricity. It is true that Thomas A. Edison has helped the progress of the world by many other inventions and discoveries quite outside of electricity, but it is in this field that he is best known. Now, in this age of electricity, it happens very fortunately that a close personal association with Mr. Edison makes it possible at last to tell younger readers the real story of Mr. Edison's life, partly in his own words. It has been a life full of surprises as well as of great achievements, and one of the surprises which we meet at the start is that, unlike Mozart, who showed his musical genius in infancy, and unlike others devoted to one thing from the outset, Edison took up electricity almost by accident.

Yet this is not so strange when we think how little electricity there was to take up in the middle of the nineteenth century. Electricity was not studied in the schools. It was not a separate art or business. Men of science had occupied themselves with electricity for a long time, but they really did not know as much about it as a bright boy in the upper grammar grades to-day. Speaking in a very general way, we may say that simple frictional electricity [1] was an old story, that Franklin had discovered the identity of electricity and lightning, and that Galvani had discovered in 1790 and Volta had developed in 1801 the generating of electric currents from batteries composed of zinc and copper plates immersed in sulphuric acid.

But it was not until 1835, only twelve years before Edison was born, that Samuel F. B. Morse applied electrical currents to the sending of an alphabet of dots and dashes by wire. Thus it was in the infancy of telegraphy that Edison first saw the light.

Telegraph apparatus in those early days was of a crude and cumbersome kind—quite different from that which young students experiment with at the present time. For instance, the receiving magnets of the earliest telegraphs, which performed the same office as the modern sounders, weighed seventy-five pounds instead of a few ounces.

It was a very difficult undertaking for Morse to establish the telegraph after he had invented it. It was such a new idea that the public could not seem to understand its use and possibilities. People would not believe that it was possible to send messages regularly over a long stretch of wire, and, even if it were possible, that it would be of much use anyway. It took him a long time to raise money to put up a telegraph line between Baltimore and Washington. Before this, he had offered to sell the whole invention outright to the United States Government for one hundred thousand dollars; but the Government did not buy, as the invention was not thought to be worth that much money.

In 1847, the year Edison was born, there were only a few telegraph circuits in existence. The farthest line to the west was in Pittsburgh, Pennsylvania. It was in this early telegraph office that Andrew Carnegie was a messenger boy. We could name a great many more notable men in our country who began their careers in a similar way, or as telegraph operators, in the early days of telegraphy, but space forbids.

Within a few years after Edison was born there came a great boom in telegraphy, and new lines were put up all over the country. Thus, by the time he had grown to boyhood the telegraph was a well-established business, and the first great electrical industry became a pronounced success.

There were no other electrical industries at this time, except electro-plating to a limited extent. The chief reason of this was probably that the only means of obtaining electrical current was by means of chemical batteries, as mechanical generators had not been developed at that time.

While the principles of the dynamo-electric machine had been discovered, and a few of these machines and small electric motors had been made by scientists, in the middle of the nineteenth century such machines were little more than scientific toys, and not to be compared with the generators of modern days.

Edison, therefore, was born at the very beginning of "The Age of Electricity," which can be said to have actually begun about 1840, or soon after.

It is not too much to say that the many important and practical inventions that he has since contributed to the electrical arts have had no small weight in causing the present time to be known as "The Age of Electricity."

[1] Made by rubbing certain objects together, like amber and silk, the original discovery over two thousand years ago.

II

EDISON'S FAMILY

Had there not been a family difference of opinion about the War of Independence, we might never have had Edison the great inventor.

The first Edisons in this country came over from Holland about the year 1730. They were descendants of a family of millers on the Zuyder Zee, and when they came to America they first settled near Caldwell, New Jersey.

Later on they removed to some land along the Passaic River. It is a curious and interesting coincidence that a hundred and sixty years later Mr. Edison established the home he now occupies in the Orange Mountains, which is in the same general neighborhood.

The family must have gotten along well in the world, for we find the name of Thomas Edison, as a bank official on Manhattan Island, signed to Continental currency in 1778. This was Mr. Edison's great-grandfather, who lived to be one hundred and four years of age.

It will be seen from the date, 1778, that this was during the time of the War of Independence. This Thomas Edison was a stanch patriot, who thoroughly believed in American independence. He had a son named John, who differed with his father in political principles and favored a continuance of British rule.

After the war was over John left the country, and, with many other Loyalists, emigrated to Nova Scotia and settled there. While he still lived there a son was born to him, at Digby, in 1804. This son was named Samuel, who became the father of Thomas Alva Edison, the inventor.

Seven years later John Edison, as a Loyalist, became entitled under the laws of Canada to a grant of six hundred acres of land, and moved westward with his family to take possession of it. He made his way through the State of New York in wagons drawn by oxen to the township of Bayfield, in upper Canada, on Lake Huron, and there settled down.

Some time afterward John Edison moved from Bayfield to Vienna, Ontario, on the northern bank of Lake Erie. As will be understood from the above, he was the grandfather of Mr. Edison, who gives this recollection of the old man in those early Canadian days:

"When I was five years old I was taken by my father and mother on a visit to Vienna. We were driven by a carriage from Milan, Ohio, to a railroad, then to a port on Lake Erie, thence by a canal-boat in a tow of several miles to Port Burwell, in Canada, across the lake, and from there we drove to Vienna, a short distance away. I remember my grandfather perfectly as he appeared at one hundred and two years of age, when he died. In the middle of the day he sat under a large tree in front of the house, facing a well-traveled road. His head was covered completely with a large quantity of very white hair, and he chewed tobacco incessantly, nodding to friends as they passed by. He used a very large cane, and walked from the chair to the house, resenting any assistance. I viewed him from a distance, and could never get very close to him. I remember some large pipes, and especially a molasses jug, a trunk, and several other things that came from Holland."

John Edison was long-lived, like his father, and died at the age of one hundred and two. Little is known of the early manhood of his son Samuel (Thomas A. Edison's father), until we find him keeping a hotel at Vienna, and in 1828 marrying Miss Nancy Elliott, who was a school-teacher there.

He was six feet in height, and was possessed of great strength and vigor. He took a lively share in the troublous politics of the period.

In 1837 the Canadian Rebellion broke out. The cause of it was the same as that which led to the War of Independence in America—taxation without representation.

Samuel Edison was so ardently interested and of such strong character that he became a captain in the insurgent forces that rallied under the banners of Papineau and Mackenzie.

The rebellion failed, however, and those who had taken part in it were severely dealt with. Many of the insurgents went in exile to Bermuda, but Samuel Edison preferred the perils of a flight to the United States. He therefore departed from Canada with his wife, hurriedly and secretly.

There was a romantic and thrilling journey of one hundred and eighty-two miles toward safety. The country through which they passed was then very wild and infested with Indians of unfriendly disposition, and the journey was made almost entirely without food or sleep.

They arrived safely in the United States, however, and, after a few years spent in various towns along the shores of Lake Erie, finally came to Milan, Ohio, in 1842. Here they settled down and made their home, for the place gave great promise of abundance of business and prosperity.

In those days railroads were few and far between, and there was none near Milan. The great quantities of grain that were grown in the surrounding country were sent to Eastern ports by sailing vessels over the lake. Milan was connected by a wide canal with the Huron River, which emptied into Lake Erie. Thus the town became a busy port, with grain warehouses and elevators, at which as many as twenty sailing vessels were loaded in a single day.

There also sprang up a brisk ship-building industry, for which the abundant forests of the region supplied the necessary lumber.

You will see, therefore, that Mr. Edison's father gave evidence of shrewd judgment when he decided to make his permanent home at Milan, for there was plenty of occupation, with every prospect of prosperity. He was always ready to look on the brightest side of everything, and could and did turn his hand to many occupations.

He decided to make his chief business the manufacture of shingles, for which there was a large demand, both in the neighborhood and along the shores of the lake. The shingles were made mostly of Canadian wood, which was imported for the purpose. They were made entirely by hand and of first-class wood, and so well did they last that a house in Milan on which these shingles were put in 1844 was still in excellent condition forty-two years later. Samuel Edison did well in this business and employed a number of men.

In a few years after the family had made their home at Milan, Thomas Alva Edison was born there, on February 11, 1847.

His mother was an attractive and highly educated woman, and her influence upon his disposition has been profound and lasting. She was born in Chenango County, New

York, in 1810, and was the daughter of the Rev. John Elliott, a Baptist minister, and descendant of an old Revolutionary soldier, Capt. Ebenezer Elliott, of Scotch descent.

The Elliott family was evidently one of considerable culture and deep religious feeling, for two of Mrs. Elliott's uncles and two brothers were also in the Baptist ministry. As a young woman she became a teacher in the public high school at Vienna, Ontario, and thus met her husband, who was residing there.

The Edison family consisted of three children, two boys and a girl. Besides Thomas Alva, there was an elder brother, William Pitt, and a sister named Tannie. Both brother and sister had considerable ability, although in different lines. William Pitt Edison was clever with his pencil, and there was at one time an idea of having him become an art student; but evidently the notion was not carried out, for later in life he was manager of the local street-railway lines at Port Huron, Michigan, in which he was heavily interested.

This talent for sketching seems to run in the family, for Thomas A. Edison's first impulse in discussing any mechanical question is to take up the nearest piece of paper and make drawings. Scarcely a day passes that this does not happen. His immense number of note-books contain thousands of such sketches.

His sister, who in later life became Mrs. Tannie Edison Bailey, had, on the other hand, a great deal of literary ability, and spent much of her time in writing.

As a child the great inventor was not at all strong, and was of fragile appearance. His head was well shaped but very large, and it is said that local doctors feared he might have brain trouble.

On account of his supposed delicacy, he was not allowed to go to school at as early an age as is usual. And when he did go, it was not for a long time. He was usually at the foot of his class, and the teacher had spoken of the boy to a school inspector as being "addled."

Perhaps the reader can imagine the indignation of his mother on hearing of this teacher's report. She had watched and studied her boy closely, and knew that he had a mind unusually receptive and mental powers far beyond those of other children. So she resolved to take him out of school and educate him herself.

It was fortunate that Mr. Edison had a mother who was not only loving, observing, and wise, but at the same time well informed and ambitious. From her experience as a teacher, she was able to give him an education better than could be had in the local schools of that day.

Under her care the boy formed studious habits and a taste for good literature that have lasted to this day. He is a great reader, and what has once been read by him is never forgotten if it is in any way useful.

When Edison was a child he was deeply interested in the busy scenes of the canal and grain warehouses, and particularly in the ship-building yards.

He asked so many questions that he fairly tired out his father, although the older man had no small ability. It has been reported that other members of the family regarded the boy as being mentally unbalanced and likely to be a lifelong care to his parents.

Even while he was quite a young child his mechanical tendencies showed themselves in his fondness for building little plank roads from the pieces of wood thrown out by the ship-building yards and the sawmills. One day he was found in the village square laboriously copying the signs of the stores.

To this day Mr. Edison is not inclined to accept a statement unless he can prove it for himself by experiment. Once, when he was about six years old, he watched a goose sitting on her eggs and saw them hatch. Soon after he was missing. By and by, after an anxious search, his father found him sitting in a nest he had made in the barn filled with goose and hen eggs he had collected, trying to hatch them out.

His remarkable memory was noticeable even when he was a child, for before he was five years old he had learned all the songs of the lumber gangs and of the canal men. Even now his recollection goes back to 1850, when, as a child three or four years old, he saw camped in front of his home six covered wagons, "prairie schooners," and witnessed their departure for California, where gold had just been discovered.

Another of his recollections of childhood is of a sadder nature. He went off one day with another boy to bathe in the creek. Soon after they entered the water the other boy disappeared. Young Edison waited around for about half an hour, and then, as it was growing dark, went home, puzzled and lonely, but said nothing about the matter. About two hours afterward, when the missing boy was being searched for, a man came to the Edison home to make anxious inquiry of the companion with whom he had last been seen. Edison told all the circumstances with a painful sense of being in some way guilty. The creek was at once dragged, and then the body was recovered.

Edison himself had more than one narrow escape. Of course, he fell into the canal and was nearly drowned— few boys in Milan worth their salt omitted that performance. On another occasion he fell into a pile of wheat in a grain elevator and was almost smothered. Holding the end of a skate-strap, that another lad might cut it with an ax, he lost the top of a finger. Fire also had its peril. He built a fire in a barn, but the flames spread so rapidly that, although he escaped himself, the barn was wholly destroyed. He was publicly whipped in the village square as a warning to other youths. Equally well remembered is a dangerous encounter with a ram which attacked him while he was busily engaged digging out a bumblebee's nest near an orchard fence, and was about to butt him again when he managed to drop over on the safe side and escape. He was badly hurt and bruised, and no small quantity of arnica was needed for his wounds.

Meanwhile railroad building had been going on rapidly, and the new Columbus, Sandusky & Hocking Railroad had reached Milan and quickly deprived it of its flourishing grain trade. The town, formerly so bustling and busy, no longer offered to so active a man as Mr. Edison's father the opportunity of conducting a prosperous business, so he decided to move away. He was well-to-do, but he determined to do better elsewhere. In 1854 he and his family removed to Port Huron, Michigan, where they occupied a large Colonial house standing in the middle of an old Government fort reservation of ten acres, overlooking the St. Clair River just after it leaves Lake Huron.

The old house at Milan where Mr. Edison was born is still in existence, and is occupied at this time (1911) by Mr. S. O. Edison, a half-brother of Edison's father, and a man of much ability.

This birthplace of Edison still remains the plain, substantial brick house it was originally, one-storied, with rooms finished on the attic floor.

III

EDISON'S EARLY BOYHOOD

It was when he was about seven years old that Edison's parents moved to Port Huron, Michigan, and it was there, a few years later, that he began his active life by becoming a newsboy.

With his mother he found study easy and pleasant. The quality of the education she gave him may be judged from the fact that before he was twelve years old he had studied the usual rudiments and had read, with his mother's help, Gibbon's *Decline and Fall of the Roman Empire*, Hume's *History of England*, Sear's *History of the World*, Burton's *Anatomy of Melancholy*, and the *Dictionary of Sciences*.

They even tried to struggle through Newton's *Principia*, but the mathematics were too much for both teacher and student. To this day Edison has little personal use for arithmetic beyond that which is called "mental." He said to a friend, "I can always hire some mathematicians, but they can't hire me."

His father always encouraged his literary tastes, and paid him a small sum for each book which he mastered. Although there is no fiction in the list, Edison has all his life enjoyed it, particularly the works of such writers as Victor Hugo. Indeed, later on, when he became a telegraph operator, he was nicknamed by his associates "Victor

Hugo Edison"—possibly because of his great admiration for that writer.

When he was about eleven years old he became greatly interested in chemistry. He got a copy of Parker's *School Philosophy*, an elementary book on physics, and tried almost every experiment in it. He also experimented on his own account. It is said that he once persuaded a boy employed by the family to swallow a large quantity of Seidlitz powders in the belief that the gases penetrated would enable him to fly. The awful agonies of the victim attracted attention, and Edison's mother marked her displeasure by an application of the switch kept behind the old Seth Thomas "grandfather's clock."

It was as early as this that young Alva, or "Al," as he was called, displayed a passion for chemistry, which has never left him. He used the cellar of the house for his experiments and collected there no fewer than two hundred bottles from various places. They contained the chemicals with which he was constant experimenting, and were all marked "Poison," so that no one else would disturb them.

He soon became familiar with all the chemicals to be had at the local drug stores, for he did not believe the statements made in his books until he had tested them for himself.

Edison used such a large part of his mother's cellar for this, his first laboratory, that, becoming tired of the "mess," she once ordered him to clear out everything. The boy was so much distressed at this that she relented, but insisted that he must keep things under lock and key when he was not there.

Most of his spare time was spent in the cellar, for he did not share to any extent in the sports of the boys of the neighborhood. His chum and chief companion at this time was a Dutch boy, much older than himself, named Michael Oates, who did chores around the house. It was Michael upon whom the Seidlitz powder experiment was tried.

As Edison got deeper into his chemical studies his limited pocket-money disappeared rapidly. He was being educated by his mother, and, therefore, not attending a regular school, and he had read all the books within reach. So he thought the matter out and decided that if he became a train newsboy he could earn all the money he wanted for his experiments and also get fresh reading from papers and magazines. Besides, if he could get permission to go on the train he had in mind, he would have some leisure hours in Detroit and would be able to spend them at the public library free of charge. His parents objected, particularly his mother, but finally he obtained their consent.

It has been thought by many people that his family was poor, and that it was on account of their poverty that young Edison came to sell newspapers on the train. This is not true, for his father was a prosperous dealer in grain and feed, and was also actively interested in the lumber industry and other things. While he was not rich, he made money in his business, and, having a well-stocked farm and a large orchard besides, was in comfortable circumstances. Socially the family stood high in the town, where at the time many well-to-do people resided.

It was of his own choice and because of his never-satisfied desire for experiment and knowledge that Edison became a newsboy.

In 1859, when he was twelve years old, he applied for the privilege of selling newspapers on the trains of the Grand Trunk Railroad between Port Huron and Detroit. After a short delay the necessary permission was obtained.

Even before this he had had some business experience. His father had laid out a "market-garden" on the farm, and young Edison, at eleven years of age, and Michael Oates had worked in it pretty steadily. In the season the two boys would load up a wagon with onions, lettuce, peas, etc., and drive through the town to sell their produce. As much as $600 was turned over to Mrs. Edison in one year from this source.

Edison was industrious, but he did not take kindly to farming. He tells us about this himself:

"After a while I tired of this work. Hoeing corn in a hot sun is unattractive, and I did not wonder that boys had left the farm for the city. Soon the Grand Trunk Railroad was extended from Toronto to Port Huron, at the foot of Lake Huron, and thence to Detroit, at about the same time the War of the Rebellion broke out. By a great amount of persistence I got permission from my mother to go on the local train as newsboy. The local train from Port Huron to Detroit, a distance of sixty-three miles, left at 7 A.M. and arrived again at 9.30 P.M. After being on the train for several months, I started two stores at Port Huron—one for periodicals and the other for vegetables, butter, and berries in the season. These were attended by two boys, who shared in the profits. The periodical store I soon closed, as the boy in charge could not be trusted. The vegetable store I kept up for nearly a year. After the railroad had been opened a short time they put on an express, which left Detroit in the morning and returned in the evening. I received permission to put a newsboy on this train. Connected with this train was a car, one part for baggage and the other part for United States mail, but for a long time it was not used. Every morning I had two large baskets of vegetables from the Detroit market loaded in the mail car and sent to Port Huron, when the boy would take them to the store. They were much better than those grown locally, and sold readily. I never was asked for freight, and to this day cannot explain why, except that I was so small and industrious and the nerve to appropriate a United States mail car to do a free freight business was so monumental. However, I kept this up for a long time, and in addition bought butter from the farmers along the line and an immense amount of blackberries in the season. I bought wholesale and at a low price, and permitted the wives of the engineers and trainmen to have the benefit of the discount. After a while there was a daily immigrant train put on. This train generally had from seven to ten coaches, filled always with Norwegians, all bound for Iowa and Minnesota. On these trains I employed a boy who sold bread, tobacco, and stick candy. As the war progressed the daily newspaper sales became very profitable, and I gave up the vegetable store."

This shrewd commercial instinct, and the capacity for carrying on successfully several business undertakings at the same time, were certainly remarkable in a boy only thirteen years old. And now, having had a glimpse of Edison's very early youth, let us begin a new chapter and follow his further adventures as a newsboy on a railway train.

IV

THE YOUNG NEWSBOY

Edison's train left Port Huron at seven o'clock in the morning and arrived at Detroit in about three hours. It did not leave Detroit again until quite late in the afternoon, arriving at Port Huron about nine-thirty at night. This made a long day for the boy, but it gave him an opportunity to do just what he wanted, which was to read, to buy chemicals and apparatus, and to indulge in his favorite occupation—chemical experimentation.

The train was made up of three coaches—baggage, smoking, and ordinary passenger. The baggage-car was divided into three compartments—one for trunks and packages, one for the mail, and one for smoking.

As there was no ventilation in this smoking-compartment, no use was made of it. It was therefore turned over to young Edison, who not only kept his papers there and his stock of goods as a "candy butcher," but he also transferred to it the contents of the precious laboratory from his mother's cellar. He found plenty of leisure on the two daily runs of the train to follow up his study of chemistry.

His earnings on the train were excellent, for he often took in eight or ten dollars a day. One dollar a day always went to his mother, and, as he was thus supporting himself, he felt entitled to spend any other profit left over on chemicals and apparatus. Detroit being a large city, he could obtain a greater variety there than in his own small town. He spent a great deal of time in reading up on his favorite subject at the public library, where he could find plenty of technical books. Thus he gave up most of his time and all his money to chemistry.

He did not confine himself entirely to chemistry in his reading at the Detroit public library, but sought to gain knowledge on other subjects. It is a matter of record that in the beginning of his reading he started in with a certain section of the library and tried to read it through, shelf by shelf, regardless of subject.

Edison went along in this manner for quite a long time. When the Civil War broke out he noticed that there was a much greater demand for newspapers. He became ambitious to publish a local journal of his own. So his little laboratory in the smoking-compartment received some additions which made it also a newspaper office.

He picked up a second-hand printing-press in Detroit and bought some type. With his mechanical ability, it was not a difficult matter to learn the rudiments of the printing art, and as some of the type was kept on the train he could set it up in moments of leisure. Thus he became the compositor, pressman, editor, proprietor, publisher, and newsdealer of the *Weekly Herald*. The price was three cents a copy, or eight cents a month for regular subscribers and the circulation ran up to over four hundred copies an issue. Only one or two copies of this journal are now to be found.

It was the first newspaper in the world printed on a train in motion. It received the patronage of the famous English engineer, Stephenson, and was also noted by the *London Times*. As the production of a boy of fourteen it was certainly a clever sheet, and many people were willing subscribers, for, by the aid of the railway telegraph, Edison was often able to print late news of local importance which could not be found in regular papers, like those of Detroit.

Edison's business grew so large that he employed a boy friend to help him. There was often plenty of work for both in the early days of the war, when the news of battle caused great excitement.

In order to increase the sales of newspapers, Edison would telegraph the news ahead to the agents of stations where the train stopped and get them to put up bulletins, so that, when the stations were reached, there would usually be plenty of purchasers waiting.

He recalls in particular the sensation caused by the great battle of Shiloh, or Pittsburg Landing, in April, 1862, in which both Grant and Sherman were engaged, in which the Confederate General Johnston was killed, and in which there was a great number of men killed and wounded.

The bulletin-boards of the Detroit newspapers were surrounded by dense crowds, which read that there were about sixty thousand killed and wounded, and that the result was uncertain. Edison, in relating his experience of that day, says:

"I knew if the same excitement was shown at the various small towns along the road, and especially at Port Huron, the sale of papers would be great. I then conceived the idea of telegraphing the news ahead, went to the operator in the depot, and, on my giving him *Harper's Weekly* and some other papers for three months, he agreed to telegraph to all the stations the matter on the bulletin-board. I hurriedly copied it, and he sent it, requesting the agents to display it on the blackboards used for stating the arrival and departure of trains. I decided that, instead of the usual one hundred papers, I could sell one thousand; but not having sufficient money to purchase that number, I determined in my desperation to see the editor himself and get credit. The great paper at that time was the *Detroit Free Press*. I walked into the office marked 'Editorial' and told a young man that I wanted to see the editor on important business—important to me, anyway.

"I was taken into an office where there were two men, and I stated what I had done about telegraphing, and that I wanted a thousand papers, but only had money for three hundred, and I wanted credit. One of the men refused it, but the other told the first spokesman to let me have them. This man, I afterward learned, was Wilbur F. Storey, who subsequently founded the *Chicago Times* and became celebrated in the newspaper world. With the aid of another boy I lugged the papers to the train and started folding them. The first station, called Utica, was a small one, where I generally sold two papers. I saw a crowd ahead on the platform, and thought it was some excursion, but the moment I landed there was a rush for me; then I realized that the telegraph was a great invention. I sold thirty-five papers there. The next station was Mount Clemens, now a watering-place, but then a town of about one thousand population. I usually sold six to eight papers there. I decided that if I found a corresponding crowd there the only thing to do to correct my lack of judgment in not getting more papers was to raise the price from five cents to ten. The crowd was there, and I raised the price. At the various towns there were corresponding crowds. It had been my practice at Port Huron to jump from the train at a point about one-fourth of a mile from the station, where the train generally slackened speed. I had drawn several loads of sand to this point to jump on, and had become quite expert. The little Dutch boy with the horse met me at this point. When the wagon approached the outskirts of the town I was met by a large crowd. I then yelled: 'Twenty-five cents apiece, gentlemen! I haven't enough to go around!' I sold out, and made what to me then was an immense sum of money."

But this and similar gains of money did not increase Edison's savings, for all his spare cash was spent for new chemicals and apparatus. He had bought a copy of Fresenius's *Qualitative Analysis*, and, with his ceaseless testing and study of its advanced problems, his little laboratory on the train was now becoming crowded with additional equipment, especially as he now added electricity to his studies.

"While a newsboy on the railroad," says Edison, "I got very much interested in electricity, probably from visiting telegraph offices with a chum who had tastes similar to mine."

We have already seen that he was shrewd enough to use the telegraph to get news items for his own little journal and also to bulletin his special news of the Civil War along the line. To such a ceaseless experimenter as he was, it was only natural that electricity should come in for a share of his attention. With his knowledge of chemistry, he had no trouble in "setting up" batteries, but his difficulty lay in obtaining instruments and material for circuits.

To-day any youth who desires to experiment with telegraphy or telephony can find plenty of stores where apparatus can be bought ready made, or he can make many things himself by following the instructions in *Harper's Electricity Book for Boys*. But in Edison's boyish days it was quite different. Telegraph supplies were hard to obtain, and amateurs were usually obliged to make their own apparatus.

However, he and his chum had a line between their homes, built of common stove-pipe wire. The insulators were bottles set on nails driven into trees and short poles. The magnet wire was wound with rags for insulation, and pieces of spring brass were used for telegraph keys.

With the idea of securing current cheaply, Edison applied the little he knew about static electricity, and actually experimented with cats. He treated them vigorously as frictional machines until the animals fled in dismay, leaving their marks to remind the young inventor of his first great lesson in the relative value of sources of electrical energy. Resorting to batteries, however, the line was made to work, and the two boys exchanged messages.

EDISON WHEN ABOUT FOURTEEN OR FIFTEEN
YEARS OF AGE

Edison wanted lots of practice, and secured it in an ingenious manner. If he could have had his way he would have sat up until the small hours of the morning, but his father insisted on eleven-thirty as the proper bed-time, which left but a short interval after a long day on the train.

Now, each evening, when the boy went home with newspapers that had not been sold, his father would sit up to read them. So Edison on some excuse had his friend take the papers, but suggested to his father that he could get the news from the chum by telegraph bit by bit. The scheme interested the father, and was put into effect, the messages over the wire being written down by Edison and handed to the old gentleman to read.

This gave good practice every night until twelve or one o'clock, and was kept up for some time, until the father became willing that his son should sit up for a reasonable time. The papers were then brought home again, and the boys practised to their hearts' content, until the line was pulled down by a stray cow wandering through the orchard.

Now we come to the incident which may be regarded as turning Edison's thoughts more definitely to electricity. One August morning, in 1862, the mixed train on which he worked as newsboy was doing some shunting at Mount Clemens station. A laden box-car had been pushed out of a siding, when Edison, who was loitering about the platform, saw the little son of the station agent, Mr. J. U. Mackenzie, playing with the gravel on the main track, along which the car, without a brakeman, was rapidly approaching.

Edison dropped his papers and his cap and made a dash for the child, whom he picked up and lifted to safety without a second to spare, as the wheel struck his heel. Both were cut about the face and hands by the gravel ballast on which they fell.

The two boys were picked up by the train-hands and carried to the platform, and the grateful father, who knew and liked the rescuer, offered to teach him the art of train telegraphy and to make an operator of him. It is needless to say that the proposal was most eagerly accepted.

Edison found time for his new studies by letting one of his friends look after the newsboy work on the train for part of the trip, keeping for himself the run between Port Huron and Mount Clemens. We have already seen that he was qualified as a beginner, and, besides, he was able to take to the station a neat little set of instruments he had just finished at a gun shop in Detroit.

What with his business as newsboy, his publication of the *Weekly Herald*, his reading and chemical and electrical experiments, Edison was leading a busy life and making rapid progress, but unexpectedly there came disaster, which brought about a sudden change. One day, as the train was running swiftly over a piece of poorly laid track, there was a sudden lurch, and a stick of phosphorus was jarred from its shelf, fell to the floor and burst into flame.

The car took fire, and Edison was trying in vain to put out the blaze when the conductor rushed in with water and saved the car. On arriving at the next station the enraged conductor put the boy off with his entire outfit, including his laboratory and printing-plant.

The origin of Edison's deafness may be told in his own words: "My train was standing by the platform at Smith's Creek station. I was trying to climb into the freight car with both arms full of papers when the conductor took me by the ears and lifted me. I felt something snap inside my head, and my deafness started from that time and has ever since progressed."

"This deafness has been a great advantage to me in various ways. When in a telegraph office I could hear only the instrument directly on the table at which I sat, and, unlike the other operators, I was not bothered by the other instruments. Again, in experimenting on the telephone, I had to improve the transmitter so that I could hear it. This made the telephone commercial, as the magneto telephone receiver of Bell was too weak to be used as a transmitter commercially. It was the same with the phonograph. The great defect of that instrument was the rendering of the overtones in music and the hissing consonants in speech. I worked over one year, twenty hours a day, Sundays and all, to get the word "specie" perfectly recorded and reproduced on the phonograph. When this was done I knew that everything else could be done—which was a fact. Again, my nerves have been preserved intact. Broadway is as quiet to me as a country village is to a person with normal hearing."

But we left young Edison on the station platform, sorrowful and indignant, as the train moved off, deserting him in the midst of his beloved possessions. He was saddened, but not altogether discouraged, and after some trouble succeeded in making his way home, where he again set up his laboratory and also his printing-office. There was some objection on the part of the family, as they feared that they might also suffer from fire, but he promised not to bring in anything of a dangerous nature.

He continued to publish the *Weekly Herald*, but after a while was persuaded by a chum to change its character and publish it under the name of *Paul Pry*, making it a journal of town gossip about local people and their affairs and peculiarities.

No copies of *Paul Pry* can now be found, but it is known that its style was distinctly personal, and the weaknesses of the townspeople were discussed in it very freely and frankly by the two boys. It caused no small offense, and in one instance Edison was pitched into the St. Clair River by one of the victims whose affairs had been given such unsought publicity.

Possibly this was one of the reasons that caused Edison to give up the paper not very long afterward. He had a great liking for newspaper work, and might have continued in that field had it not been for strong influences in other directions. There is no question, however, that he was the youngest publisher and editor of his time.

V

A FEW STORIES OF EDISON'S NEWSBOY DAYS

The Grand Trunk Railroad machine shops at Port Huron had a great attraction for young Edison. The boy who was to have much to do with the evolution of the modern electric locomotive in later years was fascinated with the mechanism of the steam locomotive. Whenever he could get the chance he would ride with the engineer in the cab, and he liked nothing better than to handle the locomotive himself during the run. Edison's own account of what happened on of these trips is very laughable. He says:

"The engine was one of a number leased to the Grank Trunk by the Chicago, Burlington & Quincy. It had bright brass bands all over the woodwork, was beautifully painted, and everything was highly polished, which was the custom up to the time old Commodore Vanderbilt stopped it on his roads. It was a slow freight train. The engineer and fireman had been out all night at a dance. After running about fifteen miles they became so sleepy that they couldn't keep their eyes open, and agreed to permit me to run the engine. I took charge, reducing the speed to about twelve miles an hour, and brought the train of seven cars to her destination at the Grand Trunk junction safely. But something occurred which was very much out of the ordinary. I was greatly worried about the water, and I knew that if it got low the boiler was likely to explode. I hadn't gone twenty miles before black, damp mud blew out of the stack and covered every part of the engine, including myself. I was about to awaken the fireman to find out the cause of this, when it stopped. Then I approached a station where the fireman always went out to the cow-catcher, opened the oil-cup on the steam-chest, and poured oil in. I started to carry out the procedure, when, upon opening the oil-cup, the steam rushed out with a tremendous noise, nearly knocking me off the engine. I succeeded in closing the oil-cup and got back in the cab, and made up my mind that she would pull through without oil. I learned afterward that the engineer always shut off steam when the fireman went to oil. This point I failed to notice. My powers of observation were very much improved after this occurrence. Just before I reached the junction another outpour of black mud occurred, and the whole engine was a sight—so much so that when I pulled into the yard everybody turned to see it, laughing immoderately. I found the reason of the mud was that I carried so much water it passed over into the stack, and this washed out all the accumulated soot."

One afternoon, about a week before Christmas, the train on which Edison was a newsboy jumped the track. Four old cars with rotten sills went all to pieces, distributing figs, raisins, dates, and candies all over the track. Hating to see so much waste, the boy tried to save all he could by eating it on the spot, but, as a result, he says, "our family doctor had the time of his life with me."

Another incident, which shows free and easy railroading and Southern extravagance, is related by Edison, as follows:

"In 1860, just before the war broke out, there came to the train one afternoon in Detroit two fine-looking young men, accompanied by a colored servant. They bought tickets for Port Huron, the terminal point for the train. After leaving the junction just outside of Detroit, I brought in the evening papers. When I came opposite the two young men, one of them said, 'Boy, what have you got?' I said, 'Papers.' 'All right.' He took them and threw them out of the window, and, turning to the colored man, said, 'Nicodemus, pay this boy.' I told Nicodemus the amount, and he opened a satchel and paid me. The passengers didn't know what to make of the transaction. I returned with the illustrated papers and magazines. These were seized and thrown out of the window, and I was told to get my money of Nicodemus. I then returned with all the old magazines and novels I had not been able to sell, thinking perhaps this would be too much for them. I was small and thin, and the layer reached above my head, and was all I could possibly carry. I had prepared a list, and knew the amount in case they bit again. When I opened the door all the passengers roared with laughter. I walked right up to the young men. One asked me what I had. I said, 'Magazines and novels.' He promptly threw them out of the window, and Nicodemus settled. Then I came in with cracked hickory nuts, then popcorn balls, and, finally, molasses candy. All went out of the window. I felt like Alexander the Great!—I had no more chances! I had sold all I had. Finally I put a rope to my trunk, which was about the size of a carpenter's chest, and started to pull this from the baggage-car to the passenger-car. It was almost too much for my strength, but at last I got it in front of those men. I pulled off my coat and hat and shoes and laid them on the chest. Then the young man asked, 'What have you got, boy?' I said, 'Everything, sir, that I can spare that is for sale.' The passengers fairly jumped with laughter. Nicodemus paid me $27 for this last sale, and threw the whole out of the door in the rear of the car. These men were from the South, and I have always retained a soft spot in my heart for a Southern gentleman."

While Edison was a newsboy on the train a request came to him one day to go to the office of E. B. Ward & Co., at that time the largest owners of steamboats on the Great Lakes. The captain of their largest boat had died suddenly, and they wanted a message taken to another captain who lived about fourteen miles from Ridgeway station on the railroad. This captain had retired, taken up some lumber land, and had cleared part of it. Edison was offered fifteen dollars by Mr. Ward to go and fetch him, but as it was a wild country and would be dark, Edison stood out for twenty-five dollars, so that he could get the companionship of another lad. The terms were agreed to. Edison arrived at Ridgeway at 8.30 P.M., when it was raining and as dark as ink. Getting with difficulty another boy to volunteer, he launched out on his errand in the pitch-black night. The two boys carried lanterns, but the road was a rough path through dense forest. The country was wild, and it was quite usual to see deer, bear, and coon skins nailed up on the sides of houses to dry. Edison had read about bears, but couldn't remember whether they were day or night prowlers. The farther they went, the more afraid they became, and every stump in the forest looked like a bear. The other lad proposed seeking safety up a tree, but Edison objected on the plea that bears could climb, and that the message must be delivered that night to enable the captain to catch the morning train. First one lantern went out, then the other. Edison says: "We leaned up against a tree and cried. I thought if I ever got out of that scrape alive I would know more about the habits of animals and everything else, and be prepared for all kinds of mischance when I again undertook an enterprise. However, the intense darkness dilated the pupils of our eyes so as to make them very sensitive, and we could just see at times the outline of the road. Finally, just as a faint gleam of daylight arrived, we entered the captain's yard and delivered the message. In my whole life I never spent such a night of horror as that, but I got a good lesson."

An amusing incident of this period is told by Edison. "When I was a boy," he says, "the Prince of Wales, the late King Edward, came to Canada (1860). Great preparations were made at Sarnia, the Canadian town opposite Port

Huron. About every boy, including myself, went over to see the affair. The town was draped in flags most profusely, and carpets were laid on the cross-walks for the Prince to walk on. There were arches, etc. A stand was built, raised above the general level, where the Prince was to be received by the Mayor. Seeing all these preparations, my idea of a prince was very high; but when he did arrive I mistook the Duke of Newcastle for him, the Duke being a fine-looking man. I soon saw that I was mistaken, that the Prince was a young stripling, and did not meet expectations. Several of us expressed our belief that a prince wasn't much after all, and said that we were thoroughly disappointed. For this one boy was whipped. Soon the Canuck boys attacked the Yankee boys, and we were all badly licked. I, myself, got a black eye. That has always prejudiced me against that kind of ceremonial and folly."

Many years afterward, when Edison had won fame by many inventions, including his electric-light system, and had been awarded the Albert Gold Medal by the Royal Society of Arts, it was this same prince who wrote a graceful letter which accompanied the medal.

Here is another of Mr. Edison's stories:

"After selling papers in Port Huron, which was often not reached until about nine-thirty at night, I seldom got home before eleven or eleven-thirty. About half-way home from the station and the town, within twenty-five feet of the road, in a dense wood, was a soldiers' graveyard, where three hundred soldiers were buried, due to a cholera epidemic which took place at Fort Gratiot, near by, many years previously. At first we used to shut our eyes and run the horse past this graveyard, and if the horse stepped on a twig my heart would give a violent movement, and it is a wonder that I haven't some valvular disease of that organ. But soon this running of the horse became monotonous, and after a while all fears of graveyards absolutely disappeared from my system. I was in the condition of Sam Houston, the pioneer and founder of Texas, who, it was said, knew no fear. Houston lived some distance from the town, and generally went home late at night, having to pass through a dark cypress swamp over a corduroy road. One night, to test his alleged fearlessness, a man stationed himself behind a tree, and enveloped himself in a sheet. He confronted Houston suddenly, and Sam stopped and said: 'If you are a man, you can't hurt me. If you are a ghost, you don't want to hurt me. And if you are the devil, come home with me; I married your sister!' "

We have already seen that Edison was of an exceedingly studious nature and full of ambition to work, experiment, and hustle. The serious side of his nature did not, however, wholly prevail. He had a keen enjoyment of a joke, even as he has now, and in his boyhood days had no particular objection if it took a practical form. The following, as related by him, is one of many:

"After the breaking out of the War there was a regiment of volunteer soldiers quartered at Fort Gratiot, the reservation extending to the boundary line of our house. Nearly every night we would hear a call such as 'Corporal of the Guard No. 1.' This would be repeated from sentry to sentry, until it reached the barracks, when Corporal of the Guard No. 1 would come and see what was wanted. I and the little Dutch boy, upon returning from the town after selling our papers, thought we would take a hand at military affairs. So one night, when it was very dark, I shouted for Corporal of the Guard No. 1. The second sentry, thinking it was the terminal sentry who shouted, repeated it to the third, and so on. This brought the corporal along the half mile, only to find that he was fooled. We tried him three nights; but the third night they were watching, and caught the little Dutch boy, took him to the lock-up at the fort, and shut him up. They chased me to the house. I rushed for the cellar. In one small compartment, where there were two barrels of potatoes and a third one nearly empty, I poured these remnants into the other barrels, sat down, and pulled the empty barrel over my head, bottom up. The soldiers had awakened my father, and they were searching for me with candles and lanterns. The corporal was absolutely certain I came into the cellar, and couldn't see how I could have gotten out, and wanted to know from my father if there was no secret hiding-place. On assurance of my father, who said that there was not, he said it was most extraordinary. I was glad when they left, as I was cramped, and the potatoes that had been in the barrel were rotten and violently offensive. The next morning I was found in bed, and received a good switching on the legs from my father, the first and only one I ever received from him, although my mother kept behind the old Seth Thomas clock a switch that had the bark worn off. My mother's ideas and mine differed at times, especially when I got experimenting and mussed up things. The Dutch boy was released next morning."

It may have seemed strange to you, on reading this and the previous chapter, that a lad so young as Edison was during the newsboy period—from about twelve to fifteen years of age—should have been allowed such wide liberty. An extensive traveler for those days, going early and returning late, an experimenter in chemistry, a publisher, printer, newsdealer, amateur locomotive engineer, and what not, covered a large range of experience and action for one so youthful.

To others of the family than his mother he was accounted a strange boy, some believing him to be mentally unbalanced. His mother, however, understood that his was no ordinary mind, for she had studied him thoroughly. While she watched him closely, she allowed him the widest possible sphere of action and encouraged his ever increasing studies.

A member of the family, in talking recently with the writer, said that when any one expressed nervousness about young Edison during his absences she would say: "Al is all right. Nothing will happen to him. God is taking care of him."

VI

THE YOUNG TELEGRAPH OPERATOR

After Edison's expulsion from the train with his laboratory and belongings, his career as a newsboy came to a sudden close. But, while he felt some disappointment, he was not discouraged and was none the less busy. As we have seen, he published his local paper for a while and also continued his chemical experiments at home. In addition, he plunged deeply into the study of telegraphy under Mr. Mackenzie's tuition.

Edison took to telegraphy enthusiastically, giving to it no less than eighteen hours a day. After some months he had made such progress that he put up a telegraph line from the station to the village, about a mile distant, and opened an office in a drug store; but the business there was very light and the office was not continued long.

A little later he became the regular operator at Port Huron. The office was in the store of a Mr. M. Walker, who sold jewelry and also newspapers and periodicals. Edison was to be found at the office both day and night, and slept there.

He says: "I became quite valuable to Mr. Walker. After working all day I worked at the office nights as well, for the reason that 'press reports' came over one of the wires until 3 A.M., and I would cut in and copy it as well as I could, to become proficient more rapidly. The goal of the rural telegraph operator was to be able to take press. Mr. Walker tried to get my father to apprentice me at twenty dollars per month, but they could not agree. I then applied for a job on the Grand Trunk Railroad as a railway operator, and was given a place, nights, at Stratford Junction, Canada."

Many years afterward Mr. Walker described the boy of sixteen as engrossed intensely in his experiments and scientific reading. The telegraph office was not a busy one, but sometimes messages taken in would remain unsent while Edison was in the cellar busy on some chemical problem.

He would be seen at times reading a scientific paper and then disappearing to buy a few sundries for experiments. Returning from the drug store with his chemicals, he would not be seen again until required by his duties, or until he had found out for himself, if possible, the truth of the statement he had been reading. If wanted for his experiment, he did not hesitate to make free use of the watchmaker's tools that lay on the table in the front window. His one idea was to do quickly when he wanted to do; and this tendency is still one of his marked characteristics.

The telegrapher's position at Stratford Junction, Canada, was taken by Edison in 1863, when he was sixteen years old, and paid him twenty-five dollars per month. In speaking of it he has since remarked that there was little difference between the telegraph of that time and that of to-day. He says: "The telegraph men couldn't explain how it worked, and I was always trying to get them to do so. I think they couldn't. I remember the best explanation I got was from an old Scotch line repairer employed by the Montreal Telegraph Company, which operated the railroad wires. He said that if you had a dog like a dachshund, long enough to reach from Edinburgh to London, if you pulled his tail in Edinburgh he would bark in London. I could understand that, but I never could get it through me what went through the dog or over the wire."

Edison was ever keenly anxious to add to his stock of experimental apparatus, as an incident of this period shows: "While working at Stratford Junction," he says, "I was told by one of the freight conductors that in the freight-house at Goodrich there were several boxes of old broken-up batteries. I went there and found over eighty cells of the well-known Grove nitric-acid battery. The operator there, who was also agent, when asked by me if I could have the electrodes of each cell, which were made of sheet platinum, gave his permission readily, thinking they were of tin. I removed them all, and they amounted to several ounces in weight. Platinum even in those days was very expensive, costing several dollars an ounce, and I owned only three small strips. I was overjoyed at this acquisition, and those very strips and the reworked scrap are used to this day in my laboratory, over forty years later."

It was while he was employed as a night operator at Stratford Junction that Edison's inventiveness was first displayed. In order to make sure that the operators were not asleep they were required to send the signal "6" to the train despatcher's office every hour during the night. Now, Edison spent all day in study and experiment, but he needed sleep, just as any healthy youth does, and so he made a small wheel with notches on the rim and attached it to the clock and line. At night he connected it with the circuit, and at each hour the wheel revolved and automatically sent in the dots required for "sixing."

The invention was a success, but the train despatcher soon noticed that frequently, in spite of the regularity of the report, Edison's office could not be raised even if a message were sent immediately after. An investigation followed, which revealed this ingenious device, and he received a reprimand.

A serious occurrence that might have resulted in accident drove him soon after from Canada, although the youth could hardly be held to blame for it. Edison says: "This night job just suited me, as I could have the whole day to myself. I had the faculty of sleeping in a chair any time for a few minutes at a time. I taught the night yardman my call, so I could get half an hour's sleep now and then between trains, and in case the station was called the watchman would awaken me. One night I got an order to hold a freight train, and I replied that I would. I rushed out to find the signalman, but before I could find him and get the signal set the train ran past. I ran to the telegraph office, and reported that I could not hold her. The train despatcher, on the strength of my message that I would hold the train, had permitted another to leave the last station in the opposite direction. There was a lower station near the junction, where the day operator slept. I started for it on foot. The night was dark, and I fell into a culvert and was knocked senseless."

Fortunately, the two engineers saw each other approaching and stopped in time to prevent an accident. Edison, however, was summoned to the general manager's office to be tried for neglect of duty. During the trial two Englishmen called, and while they were talking with the manager the youthful operator slipped out, jumped on a freight train going to Sarnia, and was not happy until the ferry-boat from Sarnia had landed him safe on the Michigan shore.

The same winter, of 1863-64, while at Port Huron, Edison had a further opportunity of showing his ingenuity. An ice-jam had broken the telegraph cable laid in the bed of the river across to Sarnia, and communication was interrupted.

The river is three-quarters of a mile wide, and could not be crossed on foot, nor could the cable be repaired.

Edison suggested using the steam whistle of a locomotive to give the long and short signals of the Morse code. An operator on the Sarnia shore was quick enough to understand the meaning of the strange whistling, and thus messages were sent in wireless fashion across the ice-floes in the river.

Young Edison had no inclination to return to Canada after his late experience there. He decided, however, that he would stick to telegraphy as a business, and, after a short stay at home in Port Huron, set out to find work as an operator in another city. And thus he commenced the roaming and drifting life which in the next five years took him all over the Middle States.

At this time the Civil War was in progress, and many hundreds of skilled operators were at the front with the army, engaged exclusively in government service. Consequently there was a great scarcity of telegraphers throughout all the cities and towns of the country. For this reason it was not difficult for an operator to get work wherever he might go. Thus one might gratify a desire to travel and get experience without running much risk of privation.

There were a great many others besides Edison who wandered about from city to city, working awhile in one place and drifting to another. As a rule, they were bright, happy-go-lucky fellows, full of the spirit of good comradeship, and willing to share bed, board, and pocket-money with those who might temporarily be less fortunate than themselves.

Many of them used telegraphy as a stepping-stone to better themselves in life, while others, unfortunately, became dissipated, and, becoming unreliable through drink, could not hold a position for long. Had Edison been by nature less persistent and industrious than he was, this miscellaneous companionship might have tended to wreck his career, but all through his life, from boyhood, he has been particularly abstemious and has had a contempt for the wastefulness of time, money, and health entailed by the drink habit.

Throughout this period of his life Edison, although wandering from place to place, never ceased to study, explore, and experiment. Referring to this beginning of his career, he mentions a curious fact that throws light on his ceaseless application. "After I became a telegraph operator," he says, "I practised for a long time to become a rapid reader of print, and got so expert I could sense the meaning of a whole line at once. This faculty, I believe, should be taught in schools, as it appears to be easily acquired. Then one can read two or three books in a day, whereas if each word at a time only is sensed reading is laborious."

During this wandering period of his life Edison made many friends, one of the earliest of whom was Milton F. Adams, who had a strange career. Of him Edison says: "Adams was one of a class of operators never satisfied to work at any place for any great length of time. He had the 'wanderlust.' After enjoying hospitality in Boston in 1868-69, on the floor of my hall bedroom, which was a paradise for the entomologist, while the boarding-house itself was run on the Banting system of flesh reduction, he came to me one day and said: 'Good-by, Edison, I have got sixty cents, and I am going to San Francisco.' And he did go. How, I never knew personally. I learned afterward that he got a job there, and then within a week they had a telegraphers' strike. He got a big torch and sold patent medicine on the streets at night to support the strikers. Then he went to Peru as partner of a man who had a grizzly bear which they proposed entering against a bull in the bull-ring in that city. The grizzly was killed in five minutes, and so the scheme died. Then Adams crossed the Andes, and started a market report bureau in Buenos Ayres. This didn't pay, so he started a restaurant in Pernambuco, Brazil. There he did very well, but something went wrong (as it always does to a nomad), so he went to the Transvaal, and ran a panorama called 'Paradise Lost' in the Kaffir kraals. This didn't pay, and he became the editor of a newspaper; then he went to England to raise money for a railroad in Cape Colony. Next I heard of him in New York, having just arrived from Bogota, United States of Columbia, with a power of attorney and two thousand dollars from a native of that republic, who applied for a patent for tightening a belt to prevent it from slipping on a pulley—a device which he thought a new and great invention, but which was in use ever since machinery was invented. I gave Adams then a position as salesman for electrical apparatus. This he soon got tired of, and I lost sight of him."

VII

ADVENTURES OF A TELEGRAPH OPERATOR

The first position that Edison took after leaving Canada so hurriedly was at Adrian, Michigan, and of what happened there he tells a story typical of his wanderings for several years to come.

"After leaving my first job at Stratford Junction I got a position as operator on the Lake Shore & Michigan Southern at Adrian, Michigan, in the division superintendent's office. As usual, I took the 'night trick,' which most operators disliked, but which I preferred, as it gave me more leisure to experiment. I had obtained from the station agent a small room, and had established a little shop of my own. One day the day operator wanted to get off, and I was on duty. About nine o'clock the superintendent handed me a despatch which he said was very important, and which I must get off at once. The wire at the time was very busy, and I asked if I should break in. I got orders to do so, and, acting under those orders of the superintendent, I broke in and tried to send the despatch; but the other operator would not permit it, and the struggle continued for ten minutes. Finally I got possession of the wire and sent the message. The superintendent of telegraph, who then lived in Adrian and went to his office in Toledo every day, happened that day to be in the Western Union office up-town—and it was the superintendent I was really struggling with! In about twenty minutes he arrived, livid with rage, and I was discharged on the spot. I informed him that the general superintendent had told me to break in and send the despatch, but the general superintendent then and there repudiated the whole thing. Their families were socially close, so I was sacrificed. My faith in human nature got a slight jar."

From Adrian Edison went to Toledo, Ohio, and secured a position at Fort Wayne, on the Pittsburg, Fort Wayne & Chicago Railroad. This was a "day job," and he did not like it. Two months later he drifted to Indianapolis, arriving there in the fall of 1864, when for the first time he entered the employ of the Western Union Telegraph Company, with which in later years he entered into closer relationship. At this time, however, he was assigned to duty at Union Station, at a salary of seventy-five dollars a month.

He did not stay long in Indianapolis, however, leaving in February, 1865, and going from there to Cincinnati. This change was possibly caused by one of his early inventions, which has been spoken of by an expert as probably the most simple and ingenious arrangement of connections for a repeater.

His ambition was to take "press report," which would come over the wire quite fast, but finding even after considerable practice, that he "broke" frequently, he adjusted two embossing Morse registers—one to receive the press matter and the other to repeat the dots and dashes at a lower speed, so that the message could be copied leisurely. Hence, he could not be rushed or "broken" in receiving, while he could turn out copy that was a marvel of neatness and clearness. This went well under ordinary conditions, but when an unusual pressure occurred he fell behind, and the newspapers complained of the slowness with which the reports were delivered to them. As to this device, Mr. Edison said recently: "Together we took press for several nights, my companion keeping the apparatus in adjustment and I copying. The regular press operator would go to the theater or take a nap, only finishing the report after 1 A.M. One of the newspapers complained of bad copy toward the end of the report—that is, from 1 to 3 A.M.—and requested that the operators taking the report up to 1 A.M., which were ourselves, take it all, as the copy then was perfectly unobjectionable. This led to an investigation by the manager, and the scheme was forbidden.

"This instrument many years afterward was applied by me to transferring messages from one wire to any other wire simultaneously or after any interval of time. It consisted of a disk of paper, the indentations being formed in a volute spiral, exactly as in the disk phonograph to-day. It was this instrument which gave me the idea of the phonograph while working on the telephone."

Arriving in Cincinnati, Edison got employment in the Western Union Commercial Telegraph Department at sixty dollars per month. Here he made the acquaintance of Milton F. Adams, referred to in the preceding chapter. Speaking of that time, Mr. Adams says:

"I can well recall when Edison drifted in to take a job. He was a youth of about eighteen years, decidedly unprepossessing in dress and rather uncouth in manner. I was twenty-one, and very dudish. He was quite thin in those days, and his nose was very prominent, giving a Napoleonic look to his face, although the curious resemblance did not strike me at the time. The boys did not take to him cheerfully, and he was lonesome. I sympathized with him, and we became close companions. As an operator he had no superiors, and very few equals. Most of the time he was 'monkeying' with the batteries and circuits, and devising things to make the work of telegraphy less irksome. He also relieved the monotony of office work by fitting up the battery circuits to play jokes on his fellow-operators, and to deal with the vermin that infested the premises. He arranged in the cellar what he called his 'rat paralyzer,' a very simple contrivance, consisting of two plates insulated from each other and connected with the main battery. They were so placed that when a rat passed over them the fore feet on the one plate and the hind feet on the other completed the circuit, and the rat departed this life, electrocuted." Shortly after Edison's arrival in Cincinnati came the close of the Civil War and the assassination of President Lincoln. One of Edison's reminiscences is interesting as showing the mechanical way in which some telegraph operators do their work. "I noticed," he says, "an immense crowd gathering in the street outside a newspaper office. I called the attention of the other operators to the crowd, and we sent a messenger boy to find the cause of the excitement. He returned in a few minutes and shouted, 'Lincoln's shot!' Instinctively the operators looked from one face to another to see which man had received the news. All the faces were blank, and every man said he had not taken a word about the shooting. 'Look over your files,' said the boss to the man handling the press stuff. For a few moments we waited in suspense, and then the man held up a sheet of paper containing a short account of the shooting of the President. The operator had worked so mechanically that he had handled the news without the slightest realization of its significance."

Edison's diversions in Cincinnati were characteristic of his life before and since. He read a great deal, but spent most of his leisure time experimenting. Occasionally he would indulge in some form of amusement, but this was not often. At this time he and Adams were close friends, and Mr. Adams remarks: "Edison and I were fond of tragedy. Forrest and John McCullough were playing at the National Theater, and when our capital was sufficient we would go to see those eminent tragedians alternate in Othello and

Iago. Edison always enjoyed Othello greatly. Aside from an occasional visit to the Loewen Garten, 'over the Rhine,' with a glass of beer and a few pretzels consumed while listening to the excellent music of a German band, the theater was the sum and substance of our innocent dissipation."

While Edison was in Cincinnati there came one day a delegation of five trade-union operators from Cleveland to form a local branch in Cincinnati. The occasion was one of great conviviality. Night came and many of the operators were away. The Cleveland wire was in special need, and Edison, almost alone in the office, devoted himself to it all through the night and until three o'clock next morning, when he was relieved. He had been previously getting eighty dollars a month, and added to this by copying plays for a theater.

His rating was that of a "plug," or inferior operator, but having determined to become a first-class operator, he had kept up a practice of going to the office at night to take "press," acting willingly as a substitute for any operator who wanted to get off for a few hours—which often meant all night.

Thus he had been unconsciously preparing for the special ordeal which the conviviality of the trade-unionists had brought about.

Speaking of that night's work, Edison says: "My copy looked fine if viewed as a whole, as I could write a perfectly straight line across the wide sheet, which was not ruled. There were no flourishes, but the individual letters would not bear close inspection. When I missed understanding a word there was no time to think what it was, so I made an illegible one to fill in, trusting to the printers to sense it. I knew they could read anything, although Mr. Bloss, an editor of the *Inquirer*, made such bad copy that one of his editorials was pasted up on the notice board in the telegraph office with an offer of one dollar to any man who could 'read twenty consecutive words.' Nobody ever did it. When I got through I was too nervous to go home, and so I waited the rest of the night for the day manager, Mr. Stevens, to see what was to be the outcome of this union formation and of my efforts. He was an austere man, and I was afraid of him. I got the morning papers, which came out at 4 A.M., and the press report read perfectly, which surprised me greatly. I went to work on my regular day wire to Portsmouth, Ohio, and there was considerable excitement, but nothing was said to me, neither did Mr. Stevens examine the copy on the office hook, which I was watching with great interest. However, about 3 P.M. he went to the hook, grabbed the bunch and looked at it as a whole without examining it in detail, for which I was thankful. Then he jabbed it back on the hook, and I knew I was all right. He walked over to me, and said: 'Young man, I want you to work the Louisville wire nights; your salary will be one hundred and twenty-five dollars.' Thus I got from the plug classification to that of a 'first-class man.'"

Not long after this promotion was secured Edison started again on his wanderings. He went south, while his friend Adams went north, neither one having any difficulty in making the trip. He says: "The boys in those days had extraordinary facilities for travel. As a usual thing it was only necessary for them to board a train and tell the conductor they were operators. Then they could go as far as they liked. The number of operators was small, and they were in demand everywhere."

Edison's next stopping place was Memphis, Tennessee, where he got a position as operator. Here again he began to invent and improve on existing apparatus, with the result of being obliged once more to "move on." He tells the story as follows: "I was not the inventor of the auto-repeater, but while in Memphis I worked on one. Learning that the chief operator, who was a protégé of the superintendent, was trying in some way to put New York and New Orleans together for the first time since the close of the war, I redoubled my efforts, and at two o'clock one morning I had them speaking to each other. The office of the Memphis *Avalanche* was in the same building. The paper got wind of it and sent messages. A column came out in the morning about it; but when I went to the office in the afternoon to report for duty I was discharged without explanation. The superintendent would not even give me a pass to Nashville, so I had to pay my fare. I had so little money left that I nearly starved at Decatur, Alabama, and had to stay three days before going on north to Nashville. Arrived in that city, I went to the telegraph office, got money enough to buy a little solid food, and secured a pass to Louisville. I had a companion with me who was also out of a job. I arrived at Louisville on a bitterly cold day, with ice in the gutters. I was wearing a linen duster and was not much to look at, but got a position at once, working on a press wire. My traveling companion was less successful on account of his 'record.' They had a limit even in those days when the telegraph service was so demoralized."

After the Civil War was over the telegraph service was in desperate condition, and some of Mr. Edison's reminiscences of these times are quite interesting. He says: "The telegraph was still under military control, not having been turned over to the original owners, the Southern Telegraph Company. In addition to the regular force, there was an extra force of two or three operators, and some stranded ones, who were a burden to us, for board was high. One of these derelicts was a great source of worry to me personally. He would come in at all hours and either throw ink around or make a lot of noise. One night he built a fire in the grate and started to throw pistol cartridges into the flames. These would explode, and I was twice hit by the bullets, which left a black-and-blue mark. Another night he came in and got from some part of the building a lot of stationery with 'Confederate States' printed at the head. He was a fine operator, and wrote a beautiful hand. He would take a sheet of paper, write capital 'A,' and then take another sheet and make the 'A' differently; and so on through the alphabet, each time crumpling the paper up in his hand and throwing it on the floor. He would keep this up until the room was filled nearly flush with the table. Then he would quit.

"Everything at that time was 'wide open.' Disorganization reigned supreme. There was no head to anything. At night myself and a companion would go over to a gorgeously furnished faro-bank and get our midnight lunch. Everything was free. There were over twenty keno-rooms running. One of them that I visited was in a Baptist church, the man with the wheel being in the pulpit and the gamblers in the pews.

"While there, the manager of the telegraph office was arrested for something I never understood, and incarcerated in a military prison about half a mile from the office. The building was in plain sight from the office and four stories high. He was kept strictly incommunicado. One day, thinking he might be confined in a room facing the office, I put my arm out of the window and kept signaling dots and dashes by the movement of the arm. I tried this several times for two days. Finally he noticed it, and, putting his arm through the bars of the window, he established communication with me. He thus sent several messages to his friends, and was afterward set free."

Another curious story told by Edison concerns a fellow operator on night duty at Chattanooga Junction at the time he was at Memphis: "When it was reported that Hood was marching on Nashville, one night a Jew came into the office about eleven o'clock in great excitement, having heard the Hood rumor. He, being a large sutler, wanted to send a message to save his goods. The operator said it was impossible—that orders had been given to send no private messages. Then the Jew wanted to bribe my friend, who steadfastly refused, for the reason, as he told the Jew, that he might be court-martialed and shot. Finally the Jew got up to eight hundred dollars. The operator swore him to secrecy and sent the message. Now, there was no such order about private messages, and the Jew, finding it out, complained to Captain Van Duzer, chief of telegraphs, who investigated the matter, and while he would not discharge the operator, laid him off indefinitely. Van Duzer was so lenient that if an operator was to wait three days and then go and sit on the stoop of Van Duzer's office all day he would be taken back. But Van Duzer swore that if the operator had taken eight hundred dollars and sent the message at the regular rate, which was twenty-five cents, it would have been all right, as the Jew would be punished for trying to bribe a military operator; but when the operator took the eight hundred dollars and then sent the message deadhead he couldn't stand it, and he would never relent."

A third typical story of this period relates to a cipher message for General Thomas. Mr. Edison narrates it as follows: "When I was an operator in Cincinnati, working the Louisville wire nights for a time, one night a man over on the Pittsburg wire yelled out: 'D. I. cipher,' which meant that there was a cipher message from the War Department at Washington, and that it was coming, and he yelled out 'Louisville.' I started immediately to call up that place. It was just at the change of shift in the office. I could not get Louisville, and the cipher message began to come. It was taken by the operator on the other table, direct from the War Department. It was for General Thomas, at Nashville. I called for about twenty minutes and notified them that I could not get Louisville. I kept at it for about fifteen minutes longer, and notified them that there was still no answer from Louisville. They then notified the War Department that they could not get Louisville. Then we tried to get it by all kinds of roundabout ways, but in no case could anybody get them at that office. Soon a message came from the War Department to send immediately for the manager of the Cincinnati office. He was brought to the office and several messages were exchanged, the contents of which, of course, I did not know, but the matter appeared to be very serious, as they were afraid of General Hood, of the Confederate Army, who was then attempting to march on Nashville; and it was important that this cipher of about twelve hundred words or so should be got through immediately to General Thomas. I kept on calling up to twelve or one o'clock, but no Louisville. About one o'clock the operator at the Indianapolis office got hold of an operator who happened to come into his office, which had a wire which ran from Indianapolis to Louisville along the railroad. He arranged with this operator to get a relay of horses, and the message was sent through Indianapolis to this operator, who had engaged horses to carry the despatches to Louisville and find out the trouble, and get the despatches through without delay to General Thomas. In those days the telegraph fraternity was rather demoralized, and the discipline was very lax. It was found out a couple of days afterward that there were three night operators at Louisville. One of them had gone over to Jeffersonville and had fallen off a horse and broken his leg, and was in a hospital. By a remarkable coincidence another of the men had been stabbed in a keno-room, and was also in a hospital, while the third operator had gone to Cynthiana to see a man hanged and had got left by the train."

I think the most important line of investigation is the production of Electricity direct from carbon.

Edison

From Memphis Edison went to Louisville. Here he remained for about two years. It was while he was there that he perfected the peculiar vertical style of writing which has since been his characteristic style. He says of this form of writing, an example of which is given above: "I developed this style in Louisville while taking press reports. My wire was connected to the 'blind' side of a repeater at Cincinnati, so that if I missed a word or sentence, or if the wire worked badly, I could not break in and get the last words, because the Cincinnati man had no instrument by which he could hear me. I had to take what came. When I got the job the cable across the Ohio River at Covington, connecting with the line to Louisville, had a variable leak in it, which caused the strength of the signaling current to make violent fluctuations. I obviated this by using several relays, each with a different adjustment, working several sounders all connected with one sounding-plate. The clatter was bad, but I could read it with fair ease. When, in addition to this infernal leak, the wires north to Cleveland worked badly it required a large amount of imagination to get the sense of what was being sent. An imagination requires an appreciable time for its exercise, and as the stuff was coming at the rate of thirty-five to forty words a minute, it was very difficult to write down what was coming and imagine what wasn't coming. Hence it was necessary to become a very rapid writer, so I started to find the fastest style. I found that the vertical style, with each letter separate and without any flourishes, was the most rapid, and that, the smaller the letter, the greater the rapidity. As I took on an average from eight to fifteen columns of news report every day, it did not take long to perfect this method."

The telegraph offices of those early days were very crude as compared with the equipments of modern times. The apparatus was generally in a very poor condition, and the wiring was of a haphazard kind. The conditions during the time of the Civil War all tended to demoralization, both of operators and apparatus.

Indeed, the following story, related by Edison, illustrates the lengths to which telegraphers could go at a time when they were in so much demand: "When I took the position there was a great shortage of operators. One night, at 2 A.M., another operator and I were on duty. I was taking press report, and the other man was working the New York wire. We heard a heavy tramp, tramp, tramp on the rickety stairs. Suddenly the door was thrown open with great violence, dislodging it from one of the hinges. There appeared in the doorway one of the best operators we had, who worked daytime, and who was of a very quiet disposition except when intoxicated. He was a great friend of the manager of the office. His eyes were bloodshot and wild, and one sleeve had been torn away from his coat. Without noticing either of us, he went up to the stove and kicked it over. The stove-pipe fell, dislocated at every joint. It was half full of exceedingly fine soot, which floated out and completely filled the room. This produced a momentary respite to his labors. When the atmosphere had cleared sufficiently to see he went around and pulled every table away from the wall, piling them on top of the stove in the middle of the room. Then he proceeded to pull the switchboard away from the wall. It was held tightly by screws. He succeeded, finally, and when it gave way he fell with the board, and, striking on a table, cut himself so that he soon became covered with blood. He then went to the battery-room and knocked all the batteries off on the floor. The nitric acid soon began to combine with the plaster in the room below, which was the public receiving-room for messengers and bookkeepers. The excess acid poured through and ate up the account-books. After having finished everything to his satisfaction, he left. I told the other operators to do nothing. We would leave things just as they were, and wait until the manager came. In the meantime, as I knew all the wires coming through to the switchboard, I rigged up a temporary set of instruments so that the New York business could be cleared up, and we also got the remainder of the press matter. At seven o'clock the day men began to appear. They were told to go downstairs and await the coming of the manager. At eight o'clock he appeared, walked around, went into the battery-room, and then came to me, saying: 'Edison, who did this?' I told him that Billy L. had come in full of soda-water and invented the ruin before him. He walked back and forth about a minute, then, coming up to my table, put his fist down, and said: 'If Billy L. ever does that again I will discharge him.' It was needless to say that there were other operators who took advantage of that kind of discipline, and I had many calls at night after that, but none with such destructive effects."

Incidents such as these, together with the daily life and work of an operator, presented one aspect of life to our young operator in Louisville. But there was another, more intellectual side, in the contact afforded with journalism and its leaders, on which Mr. Edison looks back with great satisfaction. "I remember," he says, "the discussions between the celebrated poet and journalist George D. Prentice, then editor of the *Courier-Journal*, and Mr. Tyler, of the Associated Press. I believe Prentice was the father of the humorous paragraph of the American newspaper. He was poetic, highly educated, and a brilliant talker. He was very thin and small. I do not think he weighed over one hundred and twenty-five pounds. Tyler was a graduate of Harvard, and had a very clear enunciation, and, in sharp contrast to Prentice, he was a large man. After the paper had gone to press Prentice would generally come over to Tyler's office, where I heard them arguing on the immortality of the soul, etc. I asked permission of Mr. Tyler if, after finishing the press matter, I might come in and listen to the conversation, which I did many times after. One thing I never could comprehend was that Tyler had a sideboard with liquors and generally crackers. Prentice would pour out half a glass of what they call corn whisky, and would dip the crackers in it and eat them. Tyler took it *sans* food. One teaspoonful of that stuff would put me to sleep."

Mr. Edison throws also a curious side-light on the origin of the comic paragraph in the modern American newspaper, as distributed instantly throughout the country through the telegraph. "It was the practice of the press operators all over the country at that time, when a lull occurred, to start in and send jokes or stories the day men had collected; and these were copied and pasted up on the bulletin-board. Cleveland was the originating office for 'press,' which it received from New York and sent out simultaneously to Milwaukee, Chicago, Toledo, Detroit, Pittsburg, Columbus, Dayton, Cincinnati, Indianapolis, Vincennes, Terre Haute, St. Louis and Louisville. Cleveland would call first on Milwaukee and ask if he had anything. If so, he would send it, and Cleveland would repeat it to all of us. Thus any joke or story originating anywhere in that area was known the next day all over. The press men would come in and copy anything which could be published, which was about three per cent. I collected, too, quite a large scrap-book of it, but, unfortunately, I have lost it."

Edison was always a great reader, and was in the habit of buying books at auctions and second-hand stores. One day at an auction he bought twenty unbound volumes of the *North American Review* for two dollars. These he had bound and delivered at the telegraph office. One morning, about three o'clock, he started off for home at a rapid pace with ten volumes on his shoulder. Very soon he became conscious of the fact that bullets were flying around him. He stopped, and a breathless policeman came up and seized him as a suspicious character, ordering him to drop his parcel and explain matters. Opening the package, he showed the books, somewhat to the disgust of the officer, who imagined he had caught a burglar sneaking away with his booty. Edison explained that, being deaf, he had heard no challenge, and therefore had kept moving; and the policeman remarked, apologetically, it was well for Edison he was not a better shot.

Through all his travels Edison has preserved these books, and he has them now in his library at Llewelyn Park, Orange, New Jersey.

After two years at Louisville, Edison went back North as far as Detroit, but soon returned to Louisville. At this time there was a great deal of exaggerated talk and report about the sunny life and easy wealth of South America. This idea appealed especially to telegraph operators, and young Edison, with his fertile imagination, was readily inflamed with the glowing idea of these great possibilities.

Once more he threw up his work, and, with a couple of young friends, made his way to New Orleans, where they expected to catch a specially chartered steamer for Brazil.

They arrived in New Orleans just at the time of the great riot, when the city was in the hands of a mob. The government had seized the steamer for carrying troops. The young men therefore visited another shipping office to make inquiries about vessels for Brazil.

Here they got into conversation with an old Spaniard, to whom they explained their intentions. He had lived and worked in South America, and was very emphatic in advising them that the worst thing they could do was to leave the United States, whose freedom, calm, and opportunities could not be equaled anywhere on the face of the globe. Edison took the Spaniard's advice, and made his way North again. He heard later that his two companions had gone to Vera Cruz and had died there of yellow fever.

He returned to Louisville and resumed work there. He seems to have been fairly comfortable and happy at this time. He surrounded himself with books and various apparatus, and even indited a treatise on electricity.

It is well known that Edison is very studious and a great reader, but his associates sometimes felt surprised at his fund of general information. His own words throw some light upon this subject: "The second time I was in Louisville the Telegraph Company had moved into a new office, and the discipline was now good. I took the press job. In fact, I was a very poor sender, and therefore made the taking of press report a specialty. The newspaper men allowed me to come over, after the paper went to press, at 3 A.M., and get all the exchanges I wanted. These I would take home and lay at the foot of my bed. I never slept more than four or five hours, so that I would awake at nine or ten and read these papers until dinner-time. I thus kept posted, and knew from their activity every member of Congress, and what committees they were on, and all about the topical doings, as well as the prices of breadstuffs in all the primary markets. I was in a much better position than most operators to call on my imagination to supply missing words or sentences, which were frequent in those days of old, rotten wires, badly insulated, especially on stormy nights. Upon such occasions I had to supply in some cases one-fifth of the whole matter—pure guessing—but I got caught only once. There had been some kind of convention in Virginia, in which John Minor Botts was the leading figure. There was great excitement about it, and two votes had been taken in the convention on the two days. There was no doubt that the vote the next day would go a certain way. A very bad storm came up about ten o'clock, and my wire worked badly, and there was a cessation of all signals; then I made out the words 'Minor Botts.' The next was a New York item. I filled in a paragraph about the convention and how the vote had gone as I was sure it would go. But next day I learned that, instead of there being a vote, the convention had adjourned without action until the day after."

The insatiable thirst for knowledge beyond known facts again proved Edison's undoing. Operators were strictly forbidden to remove instruments or to use batteries except on extra work. This rule did not mean much to Edison, who had access to no other instruments except those of the company. "I went one night," he says, "into the battery-room to obtain some sulphuric acid for experimenting. The carboy tipped over, the acid ran out, went through to the manager's room below, and ate up his desk and all the carpet. The next morning I was summoned before him, and told that what the company wanted was operators, not experimenters. I was at liberty to take my pay and get out."

Thus he was once more thrown upon the world. He went back to Cincinnati, and began his second term there as an operator. He was again put on night duty, much to his satisfaction. He rented a room on the top floor of an office building, bought a cot and an oil-stove, a foot lathe, and some tools.

He became acquainted with Mr. Sommers, superintendent of telegraph of the Cincinnati & Indianapolis Railroad, who gave him permission to take such scrap apparatus as he might desire that was of no use to the company.

Edison and Sommers became very friendly, and were congenial in many ways. Both of them enjoyed jokes of a practical nature, and Edison relates one of them as follows: "Sommers was a very witty man," he says, "and fond of experimenting. We worked on a self-adjusting telegraph relay, which would have been very valuable if we could have got it. I soon became the possessor of a second-hand Ruhmkorff induction coil, which, although it would only give a small spark, would twist the arms and clutch the hands of a man so that he could not let go of the apparatus. One day we went down to the roundhouse of the Cincinnati & Indianapolis Railroad and connected up the long wash-tank in the room with the coil, one electrode being connected to earth. Above this wash-room was a flat roof. We bored a hole through the roof, and could see the men as they came in. The first man as he entered dipped his hands in the water. The floor, being wet, formed a circuit, and up went his hands. He tried it the second time, with the same result. He then stood against the wall with a puzzled expression. We surmised that he was waiting for somebody else to come in, which occurred shortly after, with the same result. Then they went out, and the place was soon crowded and there was considerable excitement. Various theories were broached to explain the curious phenomenon. We enjoyed the sport immensely."

The reader must remember this occurred forty years ago, when electricity was not popularly understood. Had it occurred to-day the mystery would have soon been explained.

It is interesting to note that the germ of Edison's quadruplex originated while he was at the Cincinnati office. There he became acquainted with George Ellsworth, a telegraph operator who left the regular telegraph service to become an operator for the Confederate guerilla Morgan.

"We soon became acquainted," says Edison of this period in Cincinnati, "and he wanted me to invent a secret method of sending despatches, so that an intermediate operator could not tap the wire and understand it. He said that if it could be accomplished he could sell it to the government for a large sum of money. This suited me, and I started in and succeeded in making such an instrument, which had in it the germ of my quadruplex now used throughout the world, permitting the despatch of four messages over one wire simultaneously. By the time I had succeeded in getting the apparatus to work Ellsworth suddenly disappeared. Many years afterward I used this little device again for the same purpose. At Menlo Park, New Jersey, I had my laboratory. There were several Western Union wires cut into the laboratory and used by me in experimenting at night. One day I sat near an instrument which I had left connected during the night. I soon found it was a private wire between New York and Philadelphia, and I heard among a lot of stuff a message that surprised me. A week after that I had occasion to go to New York, and, visiting the office of the lessee of the wire, I asked him if he hadn't sent such and such a message. The expression that came over his face was a sight. He asked me how I knew of such message. I told him the circumstances, and suggested that he had better cipher such communications, or put on a secret sounder. The result of the interview was that I installed for him my old

Cincinnati apparatus, which was used thereafter for many years."

Edison's second term in Cincinnati was not a very long one. After a while he left and went home to Port Huron, where he stayed a short time. He soon became tired of comparative idleness and communicated with his old friend, Milton Adams, who was then working in Boston, and whom he wished to rejoin if he could get work promptly in the East.

Edison himself gives the details of this eventful move, when he went East to grow up with the new art of electricity. "I had left Louisville the second time, and went home to see my parents. After stopping at home for some time, I got restless, and thought I would like to work in the East. Knowing that a former operator named Adams, who had worked with me in the Cincinnati office, was in Boston, I wrote him that I wanted a job there. He wrote back that if I came on immediately he could get me in the Western Union office. I had helped out the Grank Trunk Railroad telegraph people by a new device when they lost one of the two submarine cables they had across the river, making the remaining cable act just as well for their purpose as if they had two. I thought I was entitled to a pass, which they conceded, and I started for Boston. After leaving Toronto a terrific blizzard came up and the train got snowed under in a cut. After staying there twenty-four hours, the trainmen made snow-shoes of fence-rail splints and started out to find food, which they did about a half mile away. They found a roadside inn, and by means of snow-shoes all the passengers were taken to the inn. The train reached Montreal four days late. A number of the passengers and myself went to the military headquarters to testify in favor of a soldier who had been two days late in returning from a furlough, which was a serious matter with military people, I learned. We willingly did this, for this soldier was a great story-teller, and made the time pass quickly. I met here a telegraph operator named Stanton, who took me to his boarding-house, the most cheerless I have ever been in. Nobody got enough to eat; the bedclothes were too short and too thin; it was twenty-eight degrees below zero, and the washwater was frozen solid. The board was cheap, being only one dollar and fifty cents a week.

"Stanton said that the usual live-stock accompaniment of operators' boarding-houses was absent; he thought the intense cold had caused them to hibernate. Stanton, when I was working in Cincinnati, left his position and went out on the Union Pacific to work at Julesburg, which was a cattle town at that time and very tough. I remember seeing him off on the train, never expecting to meet him again. Six months afterward, while working press wire in Cincinnati, about 2 A.M., there was flung into the middle of the operating-room a large tin box. It made a report like a pistol, and we all jumped up startled. In walked Stanton. 'Gentlemen,' he said, 'I have just returned from a pleasure trip to the land beyond the Mississippi. All my wealth is contained in my metallic traveling-case, and you are welcome to it.' The case contained one paper collar. He sat down, and I noticed that he had a woolen comforter around his neck, with his coat buttoned closely. The night was intensely warm. He then opened his coat and revealed the fact that he had nothing but the bare skin. 'Gentlemen,' said he, 'you see before you an operator who has reached the limit of impecuniosity.'"

VIII

WORK AND INVENTION IN BOSTON

When Milton Adams received Edison's letter from Port Huron he at once went over to the Western Union Office and asked the manager, Mr. George F. Milliken, if he did not want a good operator from the West.

"What kind of copy does he make?" was the cautious response. Adams says: "I passed Edison's letter through the window for his inspection. Milliken read it and a look of surprise came over his countenance as he asked me if he could take it off the line like that. I said he certainly could, and that there was nobody who could stick him. Milliken said if he was that kind of an operator I could send for him; and I wrote Edison to come on, as I had a job for him in the main office of the Western Union."

On reporting to Mr. Milliken in Boston, Edison secured a "job" very quickly. As he tells the story, he says: "The manager asked me when I was ready to go to work. 'Now,' I replied. I was then told to return at 5.30 P.M., and punctually at that hour I entered the main operating-room and was introduced to the night manager. The weather being cold, and being clothed poorly, my peculiar appearance caused much mirth, and, as I afterward learned, the night operators had consulted together how they might 'put up a job on the jay from the woolly West.' I was given a pen and assigned to the New York No. 1 wire. After waiting an hour, I was told to come over to a special table and take a special report for the *Boston Herald*, the conspirators having arranged to have one of the fastest senders in New York send the despatch and 'salt' the new man. I sat down unsuspiciously at the table, and the New York man started slowly. Soon he increased his speed, to which I easily adapted my pace. This put my rival on his mettle, and he put on his best powers, which, however, were soon reached. At this point I happened to look up, and saw the operators all looking over my shoulder, with their faces shining with fun and excitement. I knew then that they were trying to put up a job on me, but kept my own counsel. The New York man then commenced to slur over his words, running them together and sticking the signals; but I had been used to this style of telegraphy in taking report, and was not in the least discomfited. Finally, when I thought the fun had gone far enough, and having about completed the special, I quietly opened the key and remarked, telegraphically, to my New York friend, 'Say, young man, change off and send with your other foot.' This broke the New York man all up, and he turned the job over to another man to finish."

Edison did not devote his whole life at this time to the routine work of a telegraph office. His insatiable desire for knowledge led him to study deeply the underlying principles of electricity that made telegraphy possible, and he was constantly experimenting to improve the apparatus he handled daily, as well as pursuing his studies in chemistry.

One day he was more than delighted to pick up a complete set of Faraday's works. Mr. Adams says that when Edison brought home these books, at 4 A.M., he read steadily until breakfast time, and then he remarked, enthusiastically, "Adams, I have got so much to do and life is so short I am going to hustle." And thereupon he started on a run for breakfast. Edison himself says: "It was in Boston I bought Faraday's works. I think I must have tried about everything in those books. His explanations were simple. He used no mathematics. He was the master experimenter. I don't think there were many copies of Faraday's works sold in those days. The only people who did anything in electricity were the telegraphers and the opticians, making simple school aparatus to demonstrate the principles."

At this time there was a number of practical investigators and electrical workers in Boston, and Edison with his congenial tastes soon became very much at home with them. He spent a great deal of time among them, and especially in the electrical workshop of the late Charles Williams, who afterward became an associate of Alexander Graham Bell.

It was in this workshop that Edison worked out into an operative model his first patented invention, a vote recorder. This forms the subject of Edison's first patent, for which application was signed on October 11, 1868, the patent itself being taken out June 1, 1869, No. 90,646.

The purpose of this particular device was to permit a vote in the National House of Representatives to be taken in a minute or so. Edison took the vote recorder to Washington and exhibited it before a committee. In recalling the circumstance, he says: "The chairman of the committee, after seeing how quickly and perfectly it worked, said: 'Young man, if there is any invention on earth that we don't want down here it is this. One of the greatest weapons in the hands of a minority to prevent bad legislation is filibustering on votes, and this instrument would prevent it.' I saw the truth of this, because as press operator I had taken miles of Congressional proceedings, and to this day an enormous amount of time is wasted during each session of the House in foolishly calling the members' names and recording, and then adding, their votes, when the whole operation could be done in almost a moment by merely pressing a particular button at each desk. For filibustering purposes, however, the present methods are most admirable."

The outcome of this exhibition was a great disappointment to the young inventor, but it proved to be a wholesome lesson, for he determined from that time forth to devote his inventive faculties only to things for which there was a real, genuine demand. We shall see later that he has ever since lived up to the decision then made.

After the above incident Edison, with increased earnestness, resumed his study of electricity, especially in its application to telegraphy. He did not neglect his chemistry, however, but indulged his tastes freely in that direction, thus laying the foundation for the remarkable chemical knowledge that enabled him later to make some of his great inventions.

He tells an amusing incident of one of his chemical experiments of this early period: "I had read in a scientific paper the method of making nitroglycerin, and was so fired by the wonderful properties it was said to possess that I determined to make some of the compound. We tested what we considered a very small quantity, but this produced such terrible and unexpected results that we became alarmed, the fact dawning upon us that we had a very large white elephant in our possession. At 6 A.M. I put the explosive into a sarsaparilla bottle, tied a string to it, wrapped it in a paper, and gently let it down into the sewer at the corner of State and Washington Streets."

The daily routine of a telegraph office and the busy hours of reading and experimenting employed Edison's time for eighteen to twenty hours a day. Life, however, was never too strenuous for him to indulge his humor, especially if it called for the exercise of some ingenuity, as shown in the following incident related by him: "The office was on the ground floor, and had been a restaurant previous to its occupation by the Western Union Telegraph Company. It was literally loaded with cockroaches, which lived between the wall and the board running around the room at the floor, and which came after the lunch. These were such a bother on my table that I pasted two strips of tin-foil on the wall at my desk, connecting one piece to the positive pole of the big battery supplying current to the wires and the negative pole to the other strip. The cockroaches moving up on the wall would pass over the strips. The moment they got their legs across both strips there was a flash of light and the cockroaches went into gas. This automatic electrocuting device got half a column in an evening paper, and attracted so much attention that the manager made me stop it." About this time an innocent use of his chemical knowledge gave Edison a narrow escape from injury which might have shortened his career. He tells the story as follows: "After being in Boston several months, working New York wire No. 1, I was requested to work the press wire, called the 'milk route,' as there were so many towns on it taking press simultaneously. New York office had reported great delays on the wire, due to operators constantly interrupting, or 'breaking,' as it was called, to have words repeated which they had failed to get; and New York claimed that Boston was one of the worst offenders. It was a rather hard position for me, for if I took the report without breaking, it would prove the previous Boston operator incompetent. The results made the operator have some hard feelings against me. He was put back on the wire, and did much better after that. It seems that the office boy was down on this man. One night he asked me if I could tell him how to fix a key so that it would not 'break,' even if the circuit-breaker was open, and also so that it could not be easily detected. I told him to jab a penful of ink on the platinum points, as there was sugar enough in it to make it sufficiently thick to hold up when the operator tried to break—the current still going through the ink, so that he could not break.

"The next night about 1 A.M. this operator, on the press wire, while I was standing near a House printer studying it, pulled out a glass insulator, then used upside down as a substitute for an ink-bottle, and threw it with great violence at me, just missing my head. It would certainly have killed me if it had not missed. The cause of the trouble was that this operator was doing the best he could not to break, but, being compelled to open his key, he found he couldn't. The press matter came right along, and he could not stop it. The office boy had put the ink in a few minutes before, when the operator had turned his head during a lull. He blamed me instinctively as the cause of the trouble. Later we became good friends. He took his meals at the same 'emaciator' that I did. His main object in life seemed to be acquiring the art of throwing up wash-pitchers and catching them without breaking them. About a third of his salary was used up in paying for pitchers."

One of the most amusing incidents of Edison's life in Boston, occurred through a request received at the Western Union office one day from the principal of a select school for young ladies. The principal desired to have some one sent up to the school to exhibit and describe the Morse telegraph to her "children."

Edison, who was always ready to earn some extra money for his experiments, and was already known as the best-informed operator in the office, accepted the task, inviting Adams to accompany him. What happened is described by Adams as follows: "We gathered up a couple of sounders, a battery, and some wire, and at the appointed time called on her to do the stunt. Her school-room was about twenty by twenty feet, not including a small platform. We rigged up the line between the two ends of the room, Edison taking the stage, while I was at the other end of the room. All being in readiness, the principal was told to bring in her children. The door opened, and in came about twenty young ladies elegantly gowned, not one of whom was under seventeen. When Edison saw them I thought he would faint. He called me on the line and asked me to come to the stage and explain the mysteries of the Morse system. I replied that I thought he was in the right place, and told him to get busy with his talk on dots and dashes. Always modest, Edison was so overcome he could hardly speak, but he managed to say finally that, as his friend, Mr. Adams, was better equipped with cheek than he was, we would change places, and he would do the demonstrating while I explained the whole thing. This caused the bevy to turn to see where the lecturer was. I went on the stage, said something, and we did some telegraphing over the line. I guess it was satisfactory; we got the money, which was the main point to us."

Edison tells the story in a similar manner, but insists that it was he who saved the situation. "I managed to say that I would work the apparatus, and Mr. Adams would make the explanations. Adams was so embarrassed that he fell over an ottoman. The girls tittered, and this increased his embarrassment until he couldn't say a word. The situation was so desperate that for a reason I never could explain I started in myself and talked and explained better than I ever did before or since. I can talk to two or three persons, but when there are more they radiate some unknown form of influence which paralyzes my vocal cords. However, I got out of this scrape, and many times afterward when I chanced with other operators to meet some of the young ladies on their way home from school they would smile and nod, much to the mystification of the operators, who were ignorant of this episode." The purchase of supplies and apparatus for his constant experiments and studies kept Edison's pocket-money at low ebb. He never had a surplus of cash, and tells this amusing story of those impecunious days:

"My friend Adams was working in the Franklin Telegraph Company, which competed with the Western Union.

Adams was laid off, and as his financial resources had reached absolute zero centigrade, I undertook to let him sleep in my hall bedroom. I generally had hall bedrooms, because they were cheap and I needed money to buy apparatus. I also had the pleasure of his genial company at the boarding-house about a mile distant, but at the sacrifice of some apparatus. One morning, as we were hastening to breakfast, we came into Tremont Row, and saw a large crowd in front of two small 'gents" furnishing goods stores. We stopped to ascertain the cause of the excitement. One store put up a paper sign in the display window which said, 'Three hundred pairs of stockings received this day, five cents a pair—no connection with the store next door.' Presently the other store put up a sign stating they had received three hundred pairs, price three cents a pair, also that they had no connection with the store next door. Nobody went in. The crowd kept increasing. Finally, when the price had reached three pairs for one cent, Adams said to me: I can't stand this any longer; give me a cent.' I gave him a cent, and he elbowed his way in; and throwing the money on the counter, the store being filled with women clerks, he said, 'Give me three pairs.' The crowd was breathless, and the girl took down a box and drew out three pairs of baby socks. 'Oh!' said Adams, 'I want men's size.' 'Well, sir, we do not permit one to pick sizes for that amount of money.' And the crowd roared, and this broke up the sales."

During Edison's first stay in Boston he began to weary of the monotonous routine of a telegraph operator's life and took steps to establish himself in an independent business. It was at this point that he began his career as an inventor.

He says: "After the vote recorder I invented a stock ticker, and started a ticker service in Boston, had thirty or forty subscribers, and operated from a room over the Gold Exchange. This was about a year after Callahan started in New York."

It has been generally supposed that Edison did not take up stock ticker work until he left Boston finally and went to New York in 1869. But the above shows that he actually started a ticker service in Boston in 1868.

The stock ticker had been invented about a year before, 1867, by E. A. Callahan, and had then been introduced into service in New York. Its success was immediate, and it became the common ambition of every operator to invent a new ticker, as there seemed to be a promise of great wealth in this direction. Edison, however, was about the only one in Boston who seems to have achieved any tangible result.

This was not by any means all the practical work he did in Boston at this time, as we learn from his own words. He says: "I also engaged in putting up private lines, upon which I used an alphabetical dial instrument for telegraphing between business establishments, a forerunner of modern telephony. This instrument was very simple and practical, and any one could work it after a few minutes' explanation. I had these instruments made at Mr. Hamblet's, who had a little shop where he was engaged in experimenting with electric clocks. Mr. Hamblet was the father and introducer in after years of the Western Union Telegraph system of time distribution. My laboratory was the headquarters for the men, and also of tools and supplies for those private lines. They were put up cheaply, as I used the roofs of houses, just as the Western Union did. It never occurred to me to ask permission from the owners; all we did was to go to the store, etc., say we were telegraph men, and wanted to go up to the wires on the roof; and permission was always granted.

"In this laboratory I had a large induction coil which I had borrowed to make some experiments with. One day I got hold of both electrodes of the coil, and it clinched my hands on them so that I couldn't let go. The battery was on a shelf. The only way I could get free was to back off and pull the coil, so that the battery wires would pull the cells off the shelf and thus break the circuit. I shut my eyes and pulled, but the nitric acid splashed all over my face and ran down my back. I rushed to a sink, which was only half big enough, and got in as well as I could and wiggled around for several minutes to permit the water to dilute the acid and stop the pain. My face and back were streaked with yellow; the skin was thoroughly oxidized. I did not go on the street by daylight for two weeks, as the appearance of my face was dreadful. The skin, however, peeled off, and new skin replaced it without any damage."

With all the practical work he was now doing, Boston seemed to be too limited a sphere, and Edison longed for the greater opportunities of New York. His friend Adams went West to continue a life of roving and adventure, but the serious-minded Edison had had more than enough of aimless roaming, and had determined to forge ahead on the lines on which he was working.

Realizing that he must look to New York to better his fortunes, Edison, deep in debt for his new inventions, but with high hope and courage, now made the next momentous step in his career.

IX

FROM POVERTY TO INDEPENDENCE

Edison came first to New York in 1868, with his early stock printer, which he tried unsuccessfully to sell. He went back to Boston, and, quite undismayed, got up a duplex telegraph. "Toward the end of my stay in Boston," he says, "I obtained a loan of money, amounting to eight hundred dollars, to build up a peculiar kind of duplex telegraph for sending two messages over a single wire simultaneously. The apparatus was built, and I left the Western Union employ and went to Rochester, New York, to test the apparatus on the lines of the Atlantic and Pacific Telegraph between that city and New York. But the assistant at the other end could not be made to understand anything, notwithstanding I had written out a very minute description of just what to do. After trying for a week I gave it up and returned to New York with but a few cents in my pocket."

No one could have been in direr poverty than Edison when the steamboat landed him in New York in 1869. He was in debt, and his few belongings in books and instruments had to be left behind. He was not far from starving.

After leaving the boat his first thought was for breakfast; but he was without money to obtain it. He walked the streets, and in passing a wholesale tea house saw a man "tasting" tea, so he went in and asked the "taster" if he might have some tea. His request was granted, and this was his first breakfast in New York.

He knew a telegraph operator in the city, and in the course of the day succeeded in finding him, but he also was out of work, and the best he could do was to lend Edison one dollar.

By this time Edison was extremely hungry, and he gave most serious consideration as to what he should buy in the way of food that would be most satisfying. He finally decided upon apple dumplings and coffee, which he obtained at Smith & McNeil's restaurant. He says he never ate anything more appetizing.

He applied to the Western Union Company for a position as operator, but as there was no immediate vacancy he was obliged to wait for an opening. Having only the remainder of the borrowed dollar, he did not want to spend it for lodging, so he got permission to stay overnight in the battery-room of the Gold Indicator Company. Thus he kept what little change he had to buy food.

This was four years after the Civil War, but its effects were felt everywhere, and notably in the depreciation of government securities and our paper money. Gold, being the standard, was regarded as much more valuable than a paper promise to pay issued by a government heavily in debt. A gold dollar, therefore, would buy much more than a paper dollar, at times a dollar and a quarter, or a dollar and a half in value. In a word, gold commanded a high premium. For several years afterward there was a great deal of speculation in the precious metal, and a "Gold Room" had been established in Wall Street, where the transactions took place. At first the prices were exhibited on a blackboard there, but before long this plan was found to be too slow for the brokers. Then Dr. S. S. Laws, vice president and presiding officer of the Gold Exchange, invented a system of indicators to be placed in the offices of brokers. These indicators were operated from a complicated transmitting instrument at the Exchange, and each one showed the fluctuations of price as transactions took place. Dr. Laws resigned from the Exchange and organized the Gold Indicator Company, which put the system into operation.

At the time when Edison took shelter at night in the battery-room of the company there were about three hundred instruments in the offices of subscribers. While waiting to hear from the Western Union, Edison spent his days studying the indicators and the complicated transmitting instrument in the office, controlled from the keyboard of the operator on the floor of the Gold Exchange.

What happened next has been the basis of many inaccurate stories, but the following is Mr. Edison's own version: "On the third day of my arrival, and while sitting in the office, the complicated general instrument for sending on all the lines, and which made a very great noise, suddenly came to a stop with a crash. Within two minutes over three hundred boys—a boy from every broker in the street—rushed up-stairs and crowded the long aisle and office, that hardly had room for one hundred, all yelling that such and such a broker's wire was out of order and to fix it at once. It was pandemonium, and the man in charge became so excited that he lost control of all the knowledge he ever had. I went to the indicator, and, having studied it thoroughly, knew where the trouble ought to be, and found it. One of the innumerable contact springs had broken off and had fallen down between the two gearwheels and stopped the instrument; but it was not very noticeable. As I went out to tell the man in charge what the matter was Dr. Laws appeared on the scene, the most excited person I had seen. He demanded of the man the cause of the trouble, but the man was speechless. I ventured to say that I knew what the trouble was, and he said, 'Fix it! Fix it! Be quick!' I removed the spring and set the contact wheels at zero; and the line, battery, and inspecting men all scattered through the financial district to set the instruments. In about two hours things were working again. Dr. Laws came in to ask my name and what I was doing. I told him, and he asked me to come to his private office the following day. His office was filled with stacks of books all relating to metaphysics and kindred matters. He asked me a great many questions about the instruments and his system, and I showed him how he could simplify things generally. He then requested that I should call next day. On arrival, he stated at once that he had decided to put me in charge of the whole plant, and that my salary would be three hundred dollars a month! This was such a violent jump from anything I had ever had before that it rather paralyzed me for a while. I thought it was too much to be lasting; but I determined to try and live up to that salary if twenty hours a day of hard work would do it. I kept this position, made many improvements, devised several stock tickers, until the Gold and Stock Telegraph Company consolidated with the Gold Indicator Company."

Certainly few changes in fortune have been more sudden and dramatic in any notable career than this which thus placed an ill-clad, unkempt, half-starved, eager lad in a position of such responsibility in days when the fluctuations in the price of gold at every instant meant fortune or ruin to thousands.

There was at this time a very active period of speculation, and not a great while afterward came the attempt of Jay Gould and his associates to corner the gold market by buying all the available supply. This brought about the panic of Black Friday, September 24, 1869.

Edison, then but twenty-two years old, was a keen observer, and his recollection of this episode is interesting.

"On Black Friday," he says, "we had a very exciting time with the indicators. The Gould and Fisk crowd had cornered gold, and had run the quotations up faster than the indicator could follow. The indicator was composed of several wheels; on the circumference of each wheel were the numerals; and one wheel had fractions. It worked in the same way as an ordinary counter; one wheel made ten revolutions, and at the tenth it advanced the adjacent wheel; and this, in its turn having gone ten revolutions, advanced the next wheel, and so on. On the morning of Black Friday the indicator was quoting one hundred and fifty premium, whereas the bids by Gould's agents in the Gold Room were one hundred and sixty-five for five millions or any part. We had a paper-weight at the transmitter (to speed it up), and by one o'clock reached the right quotation. The excitement was prodigious. New Street, as well as Broad Street, was jammed with excited people. I sat on the top of the Western Union telegraph booth to watch the surging, crazy crowd. One man came to the booth, grabbed a pencil, and attempted to write a message to Boston. The first stroke went clear off the blank; he was so excited that he had the operator write the message for him. Amid great excitement Speyer, the banker, went crazy, and it took five men to hold him; and everybody lost their heads. The Western Union operator came to me and said: 'Shake, Edison, we are O. K. We haven't got a cent.' I felt very happy because we were poor. These occasions are very enjoyable to a poor man; but they occur rarely."

Edison in those days rather liked the modest coffee-shops and mentions visiting one. "When on the New York No. 1 wire that I worked in Boston there was an operator named Jerry Borst at the other end. He was a first-class receiver and rapid sender. We made up a scheme to hold this wire, so he changed one letter of the alphabet and I soon got used to it; and finally we changed three letters. If any operator tried to receive from Borst he couldn't do it, so

Borst and I always worked together. Borst did less talking than any operator I ever knew. Never having seen him, I went, while in New York, to call upon him. I did all the talking. He would listen, stroke his beard, and say nothing. In the evening I went over to an all-night lunch-house in Printing House Square, in a basement—Oliver's. Night editors, including Horace Greeley, and Henry Raymond, of the New York *Times*, took their midnight lunch there. When I went with Borst and another operator they pointed out two or three men who were then celebrated in the newspaper world. The night was intensely hot and close. After getting our lunch and upon reaching the sidewalk, Borst opened his mouth, and said: 'That's a great place; a plate of cakes, a cup of coffee, and a Russian bath for ten cents.' This was about fifty per cent, of his conversation for two days."

The work of Edison on the gold indicator had thrown him into close relationship with Mr. Franklin L. Pope, a young telegraph engineer, and afterward a distinguished expert and technical writer. Each recognized the special ability of the other, and barely a week after Black Friday the announcement of their partnership appeared in the *Telegrapher* of October 1, 1869.

This was the first "professional card," if it may be so described, ever issued in America by a firm of electrical engineers.

In order to be near his new friend, Edison boarded with Pope at Elizabeth, New Jersey, for some time living the "strenuous life" in the performance of his duties and following up his work on telegraph printers with marked success.

In regard to this Mr. Edison says: "While with them" (Pope and J. N. Ashley) "I devised a printer to print gold quotations instead of indicating them. The lines were started, and the whole was sold out to the Gold and Stock Telegraph Company. My experimenting was all done in the small shop of a Dr. Bradley, located near the station of the Pennsylvania Railroad in Jersey City. Every night I left for Elizabeth on the 1 A.M. train, then walked half a mile to Mr. Pope's house, and up at 6 A.M. for breakfast, to catch the 7 A.M. train. This continued all winter, and many were the occasions when I was nearly frozen in the Elizabeth walk."

After the Edison and Pope printer was bought out by the Gold and Stock Telegraph Company, its president, Gen. Marshall Lefferts, requested Edison to go to work on improving the stock ticker, he, Lefferts, to furnish the money.

Edison tackled the subject enthusiastically, and as one result produced the "Universal" ticker, which came into wide-spread use in its day. This and some other inventions had a startling effect on his fortunes. Mr. Edison says: "I made a great many inventions; one was the special ticker used for many years outside of New York in the large cities. This was made exceedingly simple, as they did not have the experts we had in New York to handle anything complicated. The same ticker was used on the London Stock Exchange. After I had made a great number of inventions and obtained patents, the General seemed anxious that the matter should be closed up. One day I exhibited and worked a successful device whereby, if a ticker should get out of unison in a broker's office and commence to print wild figures, it could be brought to unison from the central station, which saved the labor of many men and much trouble to the broker. He called me into his office, and said: 'Now, young man, I want to close up the matter of your inventions. How much do you think you should receive?' I had made up my mind that, taking into consideration the time and killing pace I was working at, I should be entitled to five thousand dollars, but could get along with three thousand dollars. When the psychological moment arrived, I hadn't the nerve to name such a large sum, so I said: 'Well, General, suppose you make me an offer.' Then he said: 'How would forty thousand dollars strike you?' This caused me to come as near fainting as I ever got. I was afraid he would hear my heart beat. I managed to say that I thought it was fair. 'All right, I will have a contract drawn; come around in three days and sign it, and I will give you the money.' I arrived on time, but had been doing some considerable thinking on the subject. The sum seemed to be very large for the amount of work, for, at that time I determined the value by the time and trouble, and not by what the invention was worth to others. I thought there was something unreal about it. However, the contract was handed to me. I signed without reading it"

Edison was then handed the first check he had ever received, one for forty thousand dollars. He went down to the bank and passed the check in to the paying teller, who handed it back to him with some remarks which in his deafness he did not hear. Fancying for a moment he had been cheated, Edison went outside "to let the cold sweat evaporate."

He went back to the General, who, with his secretary, had a good laugh over the matter, and told him the check must be endorsed, and sent with him a clerk to identify him.

The ceremony of identification performed with the paying teller, who was quite merry over the incident, Edison was given the amount in bundles of small bills "until there certainly seemed to be one cubic foot." Unaware that he was the victim of a practical joke, Edison proceeded gravely to stow away the money in his overcoat pockets and all his other pockets. He then went to Newark and sat up all night with the money for fear it might be stolen. Once more he sought help next morning, when the General laughed heartily, and, telling the clerk that the joke must not be carried any further, enabled him to deposit the currency in the bank and open an account—his first bank account.

Thus in a very brief time Edison had passed from poverty to independence. Not only that, but he had made a deep impression as to his originality and ability on important people, and had brought out valuable inventions. Thus he lifted himself at one hound out of the ranks and away from the drudgery of the key.

Many young men of twenty-two would have been so dazzled by coming suddenly into possession of forty thousand dollars after a period of poverty, struggle, and hard work, that their main ideas would have been of recreation and pleasure. Not so with Edison, however. Naturally enterprising and a pioneer, this money meant to him nothing but means to an end.

He bought some machinery and opened a small shop and got work for it. Very quickly he was compelled to move to larger quarters. Nos. 10 and 12 Ward Street, Newark, New Jersey. He secured large orders from General Lefferts to build stock tickers, and employed fifty men.

As business increased he put on a night force, and was his own foreman in both shifts. Half an hour of sleep three or four times in the twenty-four hours was all he needed. His force increased to one hundred and fifty men, and, besides superintending all the work day and night, he was constantly making new inventions in the lines on which he was then working, which was chiefly stock tickers.

A glimpse at some of young Edison's first methods as a manufacturer is interesting. He says: "Nearly all my men were on piece-work, and I allowed them to make good wages, and never cut until the pay became absurdly high as they got more expert. I kept no books. I had two hooks. All the bills and accounts I owed I jabbed on one hook, and memoranda of all owed to myself I put on the other. When some of the bills fell due, and I couldn't deliver tickers to get a supply of money, I gave a note. When the notes were due a messenger came around from the bank with the note and a protest pinned to it for one dollar and twenty-five cents. Then I would go to New York and get an advance or pay the note if I had the money. This method of giving notes for my accounts and having all notes protested I kept up over two years, yet my credit was fine. Every store I traded with was always glad to furnish goods, perhaps in amazed admiration of my system of doing business, which was certainly new."

After a while Edison got a bookkeeper, whose vagaries made him look back with regret on the earlier, primitive method. "The first three months I had him go over the books to find out how much we had made. He reported three thousand dollars. I gave a supper to some of my men to celebrate this, only to be told two days afterward that he had made a mistake, and that we had lost five hundred dollars; and then a few days after that he came to me again and said he was all mixed up, and now found that we had made over seven thousand dollars." Edison changed bookkeepers, but never afterward counted anything real profit until he had paid all his debts and had the profits in the bank.

Among the men who have worked with Edison in his various shops from time to time, there have always been those who later have risen to some notable degree of prominence in the electrical arts. This early shop was no exception.

At a single bench there worked three men since rich or prominent. One was Sigmund Bergmann, for a time partner with Edison in his lighting developments in the

United States, and now head and principal owner of electrical works in Berlin, employing ten thousand men. The next man adjacent was John Kruesi, afterward engineer of the great General Electric Works at Schenectady. A third was Schuckert, who left the bench to settle up his father's little estate at Nuremberg, stayed there and founded electrical factories which became the third largest in Germany, their proprietor dying very wealthy.

"I gave them a good training as to working hours and hustling," says Edison. And this is equally true as applied to many scores of others who have worked with him.

X

A BUSY YOUNG INVENTOR

Edison had now plunged into the intensely active life that has never since ceased. Some idea of his activity may be gained from the fact that he started no fewer than three manufacturing shops in Newark during 1870-71. All of these he directed personally, besides busying himself with many of his own schemes.

Speaking of those days, he says: "Soon after starting the large shop (10 and 12 Ward Street, Newark), I rented shop-room to the inventor of a new rifle. I think it was the Berdan. In any event, it was a rifle which was subsequently adopted by the British army. The inventor employed a tool-maker who was the finest and best I had ever seen. I noticed that he worked pretty near the whole of the twenty-four hours. This kind of application I was looking for. He was getting $21.50 a week, and was also paid for overtime. I asked him if he could run the shop. 'I don't know; try me!' he said. 'All right, I will give you sixty dollars a week to run both shifts.' He went at it. His executive ability was greater than that of any other man I have yet seen. His memory was prodigious, conversation laconic, and movements rapid. He doubled the production inside three months, without materially increasing the pay-roll, by increasing the cutting speed of tools and by the use of various devices. When in need of rest he would lie down on a work-bench, sleep twenty or thirty minutes, and wake up fresh. As this was just what I could do, I naturally conceived a great pride in having such a man in charge of my work. But almost everything has trouble connected with it. He disappeared one day, and, although I sent men everywhere that it was likely he could be found, he was not discovered. After two weeks he came into the factory in a terrible condition as to clothes and face. He sat down, and, turning to me, said: 'Edison, it's no use, this is the third time; I can't stand prosperity. Put my salary back and give me a job.' I was very sorry to learn that it was whisky that spoiled such a career. I gave him an inferior job and kept him for a long time."

Those were indeed busy days, when, at one time, Edison, besides directing the work of his shops, was working on no less than forty-five separate inventions of his own. He had thus entered definitely upon that career as an inventor which has left so deep an imprint on the records of the Patent Office.

Soon after he commenced manufacturing he was engaged by the Automatic Telegraph Company, of New York, to help it out of its difficulties. An Englishman named George Little had brought over a system of automatic telegraphy which worked well on a short line, but was a failure when put upon the longer circuits, for which automatic methods are best adapted.

This principle of automatic telegraphy, briefly described, was somewhat as follows: A narrow paper ribbon was perforated with groups of holes corresponding to Morse characters. This ribbon was passed over a cylinder, and a metallic pen was so connected that it would drop into the holes as they passed. The pen and cylinder being connected with the telegraph line, a current would pass over the line whenever the pen touched the cylinder. At the other end of the line the electrical impulses passed through another metallic pen, which rested upon another ribbon of paper chemically prepared, and, through electro-chemical action, would mark dots and dashes upon the paper.

There were a great many very serious difficulties to be overcome in order to make this system practical on long lines, but Edison applied himself to the work with tremendous energy. His laboratory note-books of the period show many thousands of experiments in the three years that he was working on his problem, and during this time he also took out a long list of patents on the subject.

So successful were his efforts that with his apparatus it became possible to send and record one thousand words a minute between New York and Washington, and thirty-five hundred words a minute between New York and Philadelphia.

Later on, Edison improved this system by further inventions, by means of which the message at the receiving end was automatically printed upon the paper ribbon in Roman letters instead of dots and dashes. Thus, the paper on which the message was received could be torn off and sent out immediately to the person for whom it was intended. This saved time and expense, for under the previous system a clerk must first translate the dots and dashes into words and write it out before delivery. The apparatus worked so perfectly that three thousand words a minute were sent between New York and Philadelphia and recorded in Roman letters.

After Edison's automatic system was put into successful use in America by the Automatic Telegraph Company, an arrangement was made for a trial of the system in England, involving its probable adoption if successful. Edison went to England in 1873 to make the demonstration. He was to report there to Col. George E. Gouraud, through whom the arrangement had been made.

With one small satchel of clothes, three large boxes of instruments, and a bright fellow-telegrapher named Jack Wright, he took voyage on the *Jumping Java*, as she was humorously known, of the Cunard line. The voyage was rough, and the little *Java* justified her reputation by jumping all over the ocean. "At the table," says Edison, "there were never more than ten or twelve people. I wondered at the time how it could pay to run an ocean steamer with so few people; but when we got into calm water and could see the green fields, I was astounded to see the number of people who appeared. There were certainly two or three hundred. Only two days could I get on deck, and on one of these a gentleman had a bad scalp wound from being thrown against the iron wall of a small smoking-room erected over a freight hatch."

Arrived in London, Edison set up his apparatus at the Telegraph Street headquarters, and sent his companion to Liverpool with the instruments for that end. The condition of the test was that he was to record at the rate of one thousand words a minute, five hundred words to be sent every half hour for six hours. Edison was given a wire and batteries to operate with, but a preliminary test soon showed that he was going to fail. Both wire and batteries were poor, and one of the men detailed by the authorities to watch the test remarked quietly, in a friendly way: "You are not going to have much show. They are going to give you an old Bridgewater Canal wire that is so poor we don't work it, and a lot of 'sand batteries' at Liverpool."[1]

The situation was rather depressing to the young American, but "I thanked him," says Edison, "and hoped to reciprocate somehow. I knew I was in a hole. I had been staying at a little hotel in Covent Garden called the Hummums, and got nothing but roast beef and flounders, and my imagination was getting into a coma. What I needed was pastry. That night I found a French pastry shop in High Holborn Street and filled up. My imagination got all right. Early in the morning I saw Gouraud, stated my case, and asked if he would stand for the purchase of a powerful battery to send to Liverpool. He said 'Yes.' I went immediately to Apps, on the Strand, and asked if he had a powerful battery. He said he hadn't; that all that he had was Tyndall's Royal Institution battery, which he supposed would not serve. I saw it—one hundred cells—and getting the price—one hundred guineas—hurried to Gouraud. He said 'Go ahead.' I telegraphed to the man in Liverpool. He came on, and got the battery to Liverpool, set up and ready just two hours before the test commenced. One of the principal things that made the system a success was that the line was put to earth at the sending end through a magnet, and the extra current from this passed to the line served to sharpen the recording waves. This new battery was strong enough to pass a powerful current through the magnet without materially diminishing the strength of the current." The test under these more favorable circumstances was a success. "The record was as perfect as copper plate, and not a single remark was made in the 'time lost' column." Edison was now asked if he thought he could get a better speed through submarine cables with this system, and replied that he would like a chance to try it. For this purpose twenty-two hundred miles of cable stored under water in tanks was placed at his disposal from 8 P.M. until 6 A.M. He says: "This just suited me, as I preferred night work. I got my apparatus down and set up, and then to get a preliminary idea of what the distortion of the signal would be I sent a single dot, which should have been recorded upon my automatic paper by a mark about one thirty-second of an inch long. Instead of that it was twenty-seven feet long. If I ever had any conceit, it vanished from my boots up! I worked on this cable more than two weeks, and the best I could do was two words per minute, which was only one-seventh of what the guaranteed speed of the cable should be when laid. What I did not know at the time was that a coiled cable, owing to induction, was infinitely worse than when laid out straight, and that my speed was as good as, if not better than, the regular system, but no one told me this."

After a short stay in England Edison returned to America. He states that the automatic was finally adopted in England and used for many years; indeed, it is still in use there. But they took whatever they needed from his system, and he "has never had a cent from them."

On arriving home he resumed arduous work on many of his inventions—chiefly those relating to duplex telegraphy. This subject had interested him at various times for four or five years previously, and he now returned to it with great vigor.

Many inventors had been working on multiple transmission, and at this period a system of sending two messages in opposite directions at the same time over one wire had been invented by Joseph Stearns, and had then lately come into use.

The subject of multiple transmission gave plenty of play for ingenuity and was one that had great fascination for Edison. He worked out many plans, and in April, 1873, two applications for patents. One of these covered an invention by which not only could two messages be sent in opposite directions over one wire at the same time, but, if desired, two separate messages could be sent simultaneously *in the same direction* over a single wire. The former method was called the "duplex," and the latter the "diplex."

Duplexing was accomplished by varying the *strength* of the current, and diplexing by *also* varying the *direction* of the current. In this invention there was the germ of the quadruplex, and now Edison redoubled his efforts toward completing the latter system, for, while duplexing doubled the capacity of a line, the quadruplex would increase it four times.

He was working also on other inventions, but the quadruplex claimed most of his attention. He says: "This problem was of the most difficult and complicated kind, and I bent all my energies toward its solution. It required a peculiar effort of the mind, such as the imagining of eight different things moving simultaneously on a mental plane without anything to demonstrate their efficiency."

It is, perhaps, hardly to be wondered at that, when notified he would have to pay twelve and one-half per cent, extra if his taxes in Newark were not at once paid, he actually forgot his own name when asked for it suddenly at the City Hall, and lost his place in the line!

He succeeded, however, in inventing a successful quadruplex system by a skilful combination of the duplex and diplex with other ingenious devices. The immense value of this invention may be realized when it is stated that it has been estimated to have saved from fifteen million to twenty million dollars in the cost of line construction in America. But Mr. Edison received only a small amount for it. We will let him tell the story in his own words:

"About this time I invented the quadruplex. I wanted to interest the Western Union Telegraph Company in it, with a view of selling it, but was unsuccessful until I made an arrangement with the chief electrician of the company, so that he could be known as a joint inventor and receive a portion of the money. At that time I was very short of money, and needed it more than glory. This electrician appeared to want glory more than money, so it was an easy trade. I brought my apparatus over and was given a separate room with a marble-tiled floor—which, by the

way, was a very hard kind of floor to sleep on—and
started in putting on the finishing touches.

"After two months of very hard work I got a detail at
regular times of eight operators, and we got it working
nicely from one room to another over a wire which ran to
Albany and back. Under certain conditions of weather one
side of the quadruplex would work very shakily, and I had
not succeeded in ascertaining the cause of the trouble. On
a certain day, when there was a board meeting of the
company, I was to make an exhibition test. The day
arrived. I had picked the best operators in New York, and
they were familiar with the apparatus. I arranged that, if a
storm occurred and the bad side got shaky, they should do
the best they could and draw freely on their imaginations.
They were sending old messages. About twelve o'clock
everything went wrong, as there was a storm somewhere
near Albany, and the bad side got shaky. Mr. Orton, the
president, and William H. Vanderbilt and the other
directors came in. I had my heart trying to climb up around
my œsophagus. I was paying a sheriff five dollars a day to
withhold execution of judgment which had been entered
against me in a case which I had paid no attention to; and
if the quadruplex had not worked before the president I
knew I was to have trouble and might lose my machinery.
The New York *Times* came out next day with a full
account. I was given five thousand dollars as part payment
for the invention, which made me easy, and I expected the
whole thing would be closed up. But Mr. Orton went on an
extended tour just about that time. I had paid for all the
experiments on the quadruplex and exhausted the money,
and I was again in straits. In the meantime I had
introduced the apparatus on the lines of the company,
where it was very successful.

"At that time the general superintendent of the Western
Union was Gen. T. T. Eckert (who had been Assistant
Secretary of War with Stanton). Eckert was secretly

negotiating with Gould to leave the Western Union and take charge of the Atlantic and Pacific—Gould's company. One day Eckert called me into his office and made inquiries about money matters. I told him Mr. Orton had gone off and left me without means, and I was in straits. He told me I would never get another cent, but that he knew a man who would buy it. I told him of my arrangement with the electrician, and said I could not sell it as a whole to anybody; but if I got enough for it I would sell all my interest in any share I might have. He seemed to think his party would agree to this. I had a set of quadruplex over in my shop, 10 and 12 Ward Street, Newark, and he arranged to bring him over next evening to see the apparatus. So the next day Eckert came over with Jay Gould and introduced him to me. This was the first time I had ever seen him. I exhibited and explained the apparatus, and they departed. The next day Eckert sent for me, and I was taken up to Gould's house, which was near the Windsor Hotel, Fifth Avenue. In the basement he had an office. It was in the evening, and we went in by the servants' entrance, as Eckert probably feared that he was watched. Gould started in at once and asked me how much I wanted. I said, 'Make me an offer.' Then he said, 'I will give you thirty thousand dollars.' I said, 'I will sell any interest I may have for that money,' which was something more than I thought I could get. The next morning I went with Gould to the office of his lawyers, Sherman & Sterling, and received a check for thirty thousand dollars, with a remark by Gould that I had got the steamboat *Plymouth Rock*, as he had sold her for thirty thousand dollars, and had just received the check. There was a big fight on between Gould's company and the Western Union, and this caused litigation. The electrician, on account of the testimony involved, lost his glory. The judge never decided the case, but went crazy a few months afterward."

Mr. Gould controlled the Atlantic and Pacific Telegraph Company and was aiming to get control of the Western Union Company, and his purchase of Edison's share in the quadruplex was an important move in this direction.

Having learned of the success of Edison's automatic system, mentioned in the early part of this chapter, Mr. Gould's next move was to get control of that. It was owned by Mr. Edison and his associates of the Automatic Telegraph Company, and that company was bought by Mr. Gould under an agreement to pay four million dollars in stock. As to this, Mr. Edison says: "After this, Gould wanted me to help install the automatic system in the Atlantic and Pacific Company, of which General Eckert had been elected president, the company having bought the Automatic Telegraph Company. I did a lot of work for this company making automatic apparatus in my shop at Newark."

Unfortunately for the inventor and his associates, the terms of the contract have never been carried out. Mr. Edison remarks in regard to this: "He" (Gould) "took no pride in building up an enterprise. He was after money, and money only. Whether the company was a success or a failure mattered not to him. After he had hammered the Western Union through his opposition company and had tired out Mr. Vanderbilt, the latter retired from control, and Gould went in and consolidated his company and controlled the Western Union. He then repudiated the contract with the Automatic Telegraph people, and they never received a cent for their wires or patents, and I lost three years of very hard labor. But I never had any grudge against him, because he was so able in his line, and as long as my part was successful the money with me was a secondary consideration. When Gould got the Western Union I knew no further progress in telegraphy was possible, and I went into other lines."

One of the most remarkable suits in the history of American jurisprudence arose out of this transaction. Mr. Edison and his associates sued Mr. Gould in 1876 for the recovery of the contract price of these inventions, and, at this writing, thirty-five years later, the suit has not been finally decided. It is now on appeal to the United States Supreme Court.

A busier shop than that of the young inventor during the years 1870 to 1874 would be difficult to find. Not only was he and it engaged on the tremendous problems of the automatic and quadruplex systems, but the shop was also busy making stock tickers. The hours were endless; and on one occasion when an order was on hand for a large quantity of these instruments Edison locked the men in until the job had been finished of making the machine perfect, and "all the bugs taken out," which meant sixty hours of hard work before the difficulties were overcome.

In addition to all this work, Edison gave attention to many other things. One of them was the first typewriter. In the early 'seventies Mr. D. N. Craig, who was interested in the automatic, brought with him from Milwaukee a Mr. Sholes, who had a wooden model of a machine to which had been given the then new and unfamiliar name of "typewriter." Mr. Craig was interested in the machine and put the model in Edison's hands to perfect.

"This typewriter proved a difficult thing," says Edison, "to make commercial. The alignment of the letters was awful. One letter would be one-sixteenth of an inch above the others, and all the letters wanted to wander out of line. I worked on it till the machine gave fair results. Some were made and used in the office of the Automatic Company. Craig was very sanguine that some day all business letters would be written on a typewriter. He died before that took place; but it gradually made its way. The typewriter I got into commercial shape is now known as the Remington. I now had five shops, and with experimenting on this new scheme I was pretty busy—at least I did not have ennui."

Later on, after the automatic was completed, and Edison was installing the system for the Atlantic and Pacific Telegraph Company he says: "About this time I invented a district messenger call-box system, and organized a company called the Domestic Telegraph Company, and started in to install the system in New York. I had great difficulty in getting subscribers, having tried several canvassers, who, one after the other, failed to get subscribers. When I was about to give it up a test operator named Brown, who was on the Automatic Telegraph wire between New York and Washington, which passed through my Newark shop, asked permission to let him try and see if he couldn't get subscribers. I had very little faith in his ability to get any, but I thought I would give him a chance, as he felt certain of his ability to succeed. He started in, and the results were surprising. Within a month he had procured two hundred subscribers, and the company was a success. I have never quite understood why six men should fail absolutely, while the seventh man should succeed. Perhaps hypnotism would account for it. This company was sold out to the Atlantic and Pacific Company."

This was not the first time that Edison had worked on district messenger signal boxes, for as far back as 1872 he had applied for a patent on a device of this kind. Although he was not the first, he was a very early inventor in this field.

It will be seen, therefore, that not all of his problems and inventions were connected with telegraphy. He seemed to find relief in working on several lines that were quite different and distinct, but all were useful and capable of wide application. For instance, when we take a piece of paraffin paper off candy, chocolate, chewing-gum or other articles, we scarcely realize that it owes its introduction to Mr. Edison. Yet such is the fact, and we relate it in his own modest words: "Toward the latter part of 1875, in the Newark shop, I invented a device for multiplying copies of letters, which I sold to Mr. A. B. Dick, of Chicago, and in the years since it has been introduced universally throughout the world. It is called the mimeograph. I also invented devices for making, and introduced, paraffin paper, now used universally for wrapping up candy, etc."

In the mimeograph a stencil is prepared by writing with a pointed pencil-like stylus on a tough prepared paper placed on a finely grooved steel plate. The pressure of the stylus causes the letters to be punctured in the sheet by a series of minute perforations, thus forming a stencil from which hundreds of copies can be made.

Edison accomplished the same perforating result by two other inventions, one a pneumatic and the other an electric motor. The latter was the one which came into extensive use, and was called the "Edison electric pen." A tiny electric motor was mounted on a pencil-like tube in which a pointed stylus (connected to the motor) traveled to and fro at a very high rate of speed. Current from a battery was supplied to the motor through a flexible cord, and the tube was held and used like a pencil, as in the other case. As many as three thousand copies have been made from such a stencil.

[1] The sand battery is now obsolete. In this type the cell containing the elements was filled with sand, which was kept moist with an electrolyte.

XI

THE TELEPHONE, MOTOGRAPH, AND MICROPHONE

It is well known that to Mr. Alexander Graham Bell belongs the credit for transmitting the articulate voice over an electric circuit by talking against a diaphragm placed in front of an electromagnet. But after Mr. Bell brought out the telephone Mr. Edison made some remarkable improvements.

In the year 1875 Edison took up the study of harmonic telegraphs, in addition to his other work, with the idea of developing a system of multiple transmission by sending sound waves over an electric circuit.

One of the devices he then made is illustrated in an interesting drawing on file at the Orange Laboratory, entitled "First Telephone on Record." This device is described by Edison in a caveat filed in the Patent Office January 14, 1876, a month before Bell filed his application for patent.

Mr. Edison states, however, that while this device was crudely capable of use as a magneto telephone, he did not invent it for transmitting speech, but as an apparatus for analyzing the complex waves arising from various sounds. He did not try the effects of sound waves produced by the human voice until after Bell's discovery was announced, but then found that this device was capable of use as a telephone.

This was a curious coincidence, but it must be understood that Mr. Edison in his testimony and public utterances has always given Mr. Bell full credit for the original discovery of transmitting articulate speech over an electric circuit.

In order to understand the value of Edison's work in this field it should be stated that, while Bell's telephone transmitted speech and other sounds, it was only practicable for short lines. Bell had no separate transmitter, but used a single apparatus both as transmitter and receiver. This instrument was similar to the receiver used to-day, having a metallic diaphragm placed near the pole of a magnet. The vibrations of the diaphragm induced very weak electric impulses in the magnetic coil. These impulses passed over the line to the receiving end, energizing the magnet coil there, and, by varying the magnetism, caused the receiving diaphragm to be similarly vibrated, and thus reproduce the sounds. Under such conditions the telephone would be practicable upon lines of only a few miles in extent, as the amount of power generated by the human voice is necessarily quite limited.

The Western Union Company requested Edison to experiment on the telephone so that it would be commercially practicable. He then went to work with a corps of helpers, and, after months of hard work day and night and the performance of many thousands of experiments, invented the carbon transmitter. This, with his plan of using an induction coil and constant battery current on the line, were the needed elements of success, and it made the telephone a commercial possibility. Every one of the many millions of telephones in use all over the world to-day bears the imprint of Edison's genius in the employment of the principles he then established.

What Edison accomplished was this: Instead of using one single apparatus for transmitting and receiving, he made a separate transmitter of special design. In this he used carbon, which varies in electrical resistance with the pressure applied. The carbon was an electrode in connection with the vibrating diaphragm, and was in a closed circuit through which flowed a battery current. The vibrations of the diaphragm caused variations of pressure on the carbon and consequent variations in the current. These in turn resulted in corresponding impulses in the receiving magnet, and the diaphragm of the receiver was vibrated accordingly, thus reproducing the sounds. Edison's plan also included the passing of the current through an induction coil, the secondary of which was connected with the main line. By this means electrical impulses of enormously high potential are sent out on the main line to the receiving end.

Thus it will be seen that with Bell's telephone the sound-waves themselves generate the electric impulses, which are hence extremely weak. With Edison's telephone the sound-waves actuate an electric valve, so to speak, and permit variations in a current of any desired strength.

Mr. Edison's own story of his telephone work is full of interest: "In 1876 I started again to experiment for the Western Union and Mr. Orton. This time it was the telephone. Bell invented the first telephone, which consisted of the present receiver, used both as a transmitter and a receiver (the magneto type). It was attempted to introduce it commercially, but it failed on account of its faintness and the extraneous sounds which came in on its wire from various causes. Mr. Orton wanted me to take hold of it and make it commercial. As I had also been working on a telegraph system employing tuning-forks, simultaneously with both Bell and Gray, I was pretty familiar with the subject. I started in, and soon produced the carbon transmitter, which is now universally used.

"Tests were made between New York and Philadelphia, also between New York and Washington, using regular Western Union wires. The noises were so great that not a word could be heard with the Bell receiver when used as a transmitter between New York and Newark, New Jersey. Mr. Orton and W. K. Vanderbilt and the board of directors witnessed and took part in the tests of my transmitter. They were successful. The Western Union then put the transmitters on private lines. Mr. Theodore Puskas, of Budapest, Hungary, was the first man to suggest a telephone exchange, and soon after exchanges were established. The telephone department was put in the hands of Hamilton McK. Twombly, Vanderbilt's ablest son-in-law, who made a success of it. The Bell Company, of Boston, also started an exchange, and the fight was on, the Western Union pirating the Bell receiver and the Boston company pirating the Western Union transmitter. About this time I wanted to be taken care of. I threw out hints of this desire. Then Mr. Orton sent for me. He had learned that inventors didn't do business by the regular process, and concluded he would close it right up. He asked me how much I wanted. I had made up my mind it was certainly worth twenty-five thousand dollars if it ever amounted to anything for central station work; so that was the sum I had in mind to obstinately stick to and get. Still it had been an easy job, and only required a few months, and I felt a little shaky and uncertain. So I asked him to make me an offer. He promptly said he would give me one hundred thousand dollars. 'All right,' I said, 'it is yours on one condition, and that is that you do not pay it all at once, but pay me at the rate of six thousand dollars a year for seventeen years—the life of the patent.' He seemed only too pleased to do this, and it was closed. My ambition was about four times too large for my business capacity, and I knew that I would soon spend this money experimenting if I got it all at once; so I fixed it that I couldn't. I saved seventeen years of worry by this stroke."

Edison continued his telephone work through a number of years and made and tested many other kinds of telephones, such as the water telephone, electrostatic telephone, condenser telephone, chemical telephone, various magneto telephones, inertia telephone, mercury telephone, voltaic pile telephone, musical transmitter, and the electromotograph.

The principle of the electromotograph was utilized by him in more ways than one; first of all in telegraphy. Soon after the time he had concluded the telephone arrangement just mentioned a patent was issued to a Mr. Page. This patent was considered very important. It related to the use of a retractile spring to withdraw the armature lever from the magnet of a telegraph or other relay or sounder, and thus controlled the art of telegraphy, except in simple circuits.

"There was no known way," remarks Edison, "whereby this patent could be evaded, and its possessor would eventually control the use of what is known as the relay and sounder, and this was vital to telegraphy. Gould was pounding the Western Union on the Stock Exchange, disturbing its railroad contracts, and, being advised by his lawyers that this patent was of great value, bought it. The moment Mr. Orton heard this he sent for me and explained the situation, and wanted me to go to work immediately and see if I couldn't evade it or discover some other means that could be used in case Gould sustained the patent. It seemed a pretty hard job, because there was no known means of moving a lever at the other end of a telegraph wire except by the use of a magnet. I said I would go at it that night. In experimenting some years previously I had discovered a very peculiar phenomenon, and that was that if a piece of metal connected to a battery was rubbed over a moistened piece of chalk resting on a metal connected to the other pole, when the current passed the friction was greatly diminished. When the current was reversed the friction was greatly increased over what it was when no current was passing. Remembering this, I substituted a piece of chalk, rotated by a small electric motor for the magnet, and connecting a sounder to a metallic finger resting on the chalk, the combination claim of Page was made worthless. A hitherto unknown means was introduced in the electric art. Two or three of the devices were made and tested by the company's expert. Mr. Orton, after he had had me sign the patent application and got it in the Patent Office, wanted to settle for it at once. He asked my price. Again I said, 'Make me an offer.' Again he named one hundred thousand dollars. I accepted, providing he would pay it at the rate of six thousand dollars a year for seventeen years. This was done, and thus, with the telephone money, I received twelve thousand dollars yearly for that period from the Western Union Telegraph Company."

A year or two later the electromotograph principle was again made use of in a curious manner. The telephone was being developed in England, and Edison had made arrangements with Colonel Gouraud, his old associate in the automatic telegraph, to represent his interests.

A company was formed, a large number of instruments were made and sent to London, and prospects were bright. Then there came a threat of litigation from the owners of the Bell patent, and Gouraud found he could not push the enterprise unless he could avoid using what was asserted to be an infringement of the Bell receiver.

He cabled for help to Edison, who sent back word telling him to hold the fort. "I had recourse again," says Edison, "to the phenomenon discovered by me some years previous, that the friction of a rubbing electrode passing over a moist chalk surface was varied by electricity. I devised a telephone receiver which was afterward known as the 'loud-speaking telephone,' or 'chalk receiver.' There was no magnet, simply a diaphragm and a cylinder of compressed chalk about the size of a thimble. A thin spring connected to the center of the diaphragm extended outwardly and rested on the chalk cylinder, and was pressed against it with a pressure equal to that which would be due to a weight of about six pounds. The chalk was rotated by hand. The volume of sound was very great. A person talking into the carbon transmitter in New York had his voice so amplified that he could be heard one thousand feet away in an open field at Menlo Park. This great excess of power was due to the fact that the latter came from the person turning the handle. The voice, instead of furnishing all the power, as with the present receiver, merely controlled the power, just as an engineer working a valve would control a powerful engine.

"I made six of these receivers and sent them in charge of an expert on the first steamer. They were welcomed and tested, and shortly afterward I shipped a hundred more. At the same time I was ordered to send twenty young men, after teaching them to become expert. I set up an exchange of ten instruments around the laboratory. I would then go out and get each one out of order in every conceivable way, cutting the wires of one, short-circuiting another, destroying the adjustment of a third, putting dirt between the electrodes of a fourth, and so on. A man would be sent to each to find out the trouble. When he could find the trouble ten consecutive times, using five minutes each, he was sent to London. About sixty men were sifted to get twenty. Before all had arrived, the Bell Company there, seeing we could not be stopped, entered into negotiations for consolidation. One day I received a cable from Gouraud offering 'thirty thousand' for my interest. I cabled back I would accept. When the draft came I was astonished to find it was for thirty thousand pounds. I had thought it was dollars."

After the consolidation of the Bell and Edison interests in England the chalk receiver was finally abandoned in favor of the Bell receiver—the latter being more simple and cheaper. Extensive litigation with newcomers into the telephone field followed, and Edison's carbon transmitter patent was sustained by the English courts, while Bell's was declared invalid.

In America, the competition between the Western Union and Bell companies, which had been keen and strenuous, was finally brought to an end under an agreement, the former company agreeing to retire from the telephonic field and the latter company agreeing to stay out of the telegraphic field. Through its ownership of Edison's carbon transmitter invention, the Western Union company came to enjoy an annual income of several hundred thousand dollars for some years as a compensation for its retirement from telephony under this agreement.

The principle involved in Edison's carbon-transmitter gave birth to another interesting device called the microphone, by means of which the faintest sounds could be very plainly heard. For instance, the footsteps of a common house-fly make a loud noise when the hearing is assisted by the microphone. As every one knows, the microphone is universally used in our modern radio.

This invention was claimed at the time for Professor Hughes, of England. Whatever credit might be due to him for the form he proposed, a standard history ascribes two original forms of the microphone to Edison, and he himself remarks: "After I sent one of my men over to London especially to show Preece the carbon transmitter, when Hughes first saw it, and heard it—then within a month he came out with the microphone, without any acknowledgment whatever. Published dates will show that Hughes came along after me."

The carbon transmitter has not been the only way in which Edison has utilized the peculiar property that carbon possesses of altering its resistance to the passage of current according to the degree of pressure brought to bear on it.

For his quadruplex system he constructed a rheostat, or resistance box, with a series of silk disks saturated with plumbago and well dried. The pressure on the disks can be regulated by an adjustable screw, and in this way the resistance of the circuit can be varied.

He also developed a "pressure," or carbon, relay, by means of which signals of variable strength can be transferred from one telegraphic circuit to another. The poles of the electromagnet in the local or relay circuit are hollowed out and filled up with carbon disks or powdered plumbago.

If a weak current passes through the relay the armature will be but feebly attracted and will only compress the carbon slightly. Thus the carbon will offer considerable resistance and the signal on the local sounder will be weak.

If, on the contrary, the incoming current be strong, the armature will be strongly attracted, the carbon will be more compressed, thus lowering the resistance and giving a loud signal on the local sounder.

Another beautiful and ingenious use of carbon was made by Edison in an instrument invented by him called the tasimeter. This device was used for indicating most minute degrees of heat, and was so exceedingly sensitive that in one case the heat of rays of light from the remote star Arcturus showed results.

The tasimeter is a very simple instrument. A strip of hard rubber rests vertically on a platinum plate, beneath which is a carbon button, under which again lies another platinum plate. The two plates and the carbon button form part of an electric circuit containing a battery and a galvanometer. Hard rubber is very sensitive to heat, and the slightest rise of temperature causes it to expand, thus increasing the pressure on the carbon button. This produces a variation in resistance shown by the swinging of the galvanometer needle.

This instrument is so sensitive that with a delicate galvanometer the heat of a person's hand thirty feet away will throw the needle off the scale.

XII

MAKING A MACHINE TALK

If one had never heard a phonograph, it would seem as though it would be impossible to take some pieces of metal and make a machine that would repeat speaking, singing, or instrumental music just like life.

So, before the autumn of 1877, when Edison invented the phonograph, the world thought such a thing was entirely out of the question. Indeed, Edison's own men in his workshop, who had seen him do some wonderful things, thought the idea was absurd when he told them that he was making a machine to reproduce human speech.

One of his men went so far as to bet him a box of cigars that the thing would be an utter failure when finished, but, as every one knows, Edison won the bet, for the very first time the machine was tried it repeated clearly all the words that were spoken into it.

A story has often been told in the newspapers that the invention was made through Edison's finger being pricked by a point attached to a vibrating telephone diaphragm, but this is not true.

The invention was not made through any accident, but was the result of pure reasoning, and in this case, as in many others, fact is more wonderful than fiction. Mr. Edison's own account of the invention of the phonograph is intensely interesting.

"I was experimenting," he says, "on an automatic method of recording telegraph messages on a disk of paper laid on a revolving platen, exactly the same as the disk talking-machine of to-day. The platen had a spiral groove on its surface, like the disk. Over this was placed a circular disk of paper; an electromagnet with the embossing point connected to an arm travelled over the disk, and any signals given through the magnets were embossed on the disk of paper. If this disk was removed from the machine and put on a similar machine provided with a contact point the embossed record would cause the signals to be repeated into another wire. The ordinary speed of telegraphic signals is thirty-five to forty words a minute; but with this machine several hundred words were possible.

"From my experiments on the telephone I knew of the power of a diaphragm to take up sound vibrations, as I had made a little toy which when you recited loudly in the funnel would work a pawl connected to the diaphragm; and this, engaging a ratchet-wheel, served to give continuous rotation to a pulley. This pulley was connected by a cord to a little paper toy representing a man sawing wood. Hence, if one shouted: 'Mary had a little lamb,' etc., the paper man would start sawing wood. I reached the conclusion that if I could record the movements of the diaphragm properly I could cause such records to reproduce the original movements imparted to the diaphragm by the voice, and thus succeed in recording and reproducing the human voice.

"Instead of using a disk I designed a little machine, using a cylinder provided with grooves around the surface. Over this was to be placed tin-foil, which easily received and recorded the movements of the diaphragm. A sketch was made, and the piece-work price, eighteen dollars, was marked on the sketch. I was in the habit of marking the price I would pay on each sketch. If the workman lost, I would pay his regular wages; if he made more than the wages, he kept it. The workman who got the sketch was John Kruesi. I didn't have much faith that it would work, expecting that I might possibly hear a word or so that would give hope of a future for the idea. Kruesi, when he had nearly finished it, asked what it was for. I told him I was going to record talking, and then have the machine talk back. He thought it absurd. However, it was finished; the foil was put on; I then shouted 'Mary had a little lamb,' etc. I adjusted the reproducer, and the machine reproduced it perfectly. I was never so taken back in my life. Everybody was astonished. I was always afraid of things that worked the first time. Long experience proved that there were great drawbacks found generally before they could be made commercial; but here was something there was no doubt of."

No wonder that John Kruesi, as he heard the little machine repeat the words that had been spoken into it, ejaculated in an awe-stricken tone: "Mein Gott im Himmel!" No wonder the "boys" joined hands and danced around Edison, singing and shouting. No wonder that Edison and his associates sat up all night fixing and adjusting it so as to get better and better results—reciting and singing and trying one another's voices and listening with awe and delight as the crude little machine repeated the words spoken or sung into it.

The news quickly became public, and the newspapers of the world published columns about this wonderful invention. Mr. Edison was besieged with letters from every part of the globe. Every one wanted to hear this machine; and in order to satisfy a universal demand for phonographs to be used for exhibition purposes he had a number of them made and turned them over to various individuals, who exhibited them to great crowds around the country. These were the machines in which the record was made on a sheet of tin-foil laid around the cylinder.

They created great excitement both in America and abroad. The announcement of a phonograph concert was sufficient to fill a hall with people who were curious to hear a machine talk and sing.

In the next year, 1878, Edison entered upon his experiments in electric lighting. His work in this field kept him intensely busy for nearly ten years, and the phonograph was laid aside as far as he was concerned.

He had not forgotten it, however, for he had fully realized its tremendous possibilities very quickly after its invention. This is shown by an article he wrote for the *North American Review*, which appeared in the summer of 1878. In that article he predicted the possible uses of the phonograph, many of which have since been fulfilled.

MR. EDISON AT THE CLOSE OF FIVE DAYS AND
NIGHTS OF CONTINUED WORK IN PERFECTING
THE EARLY WAX-CYLINDER TYPE OF
PHONOGRAPH—JUNE 16, 1888

This is the longest continuous session of labor he ever
performed.

In 1887, having finished the greatest part of his work on
the electric light, he turned to the phonograph once more.
Realizing that the tin-foil machine was not an ideal type
and could not come into common use, he determined to re-
design it, and make it an instrument that could be handled
by any one.

This meant the design and construction of an entirely different type of machine, and resulted in the kind of phonograph with which every one is familiar in these modern days. One of the chief differences was the use of a wax cylinder instead of tin-foil, and, instead of indenting with a pointed stylus, the record is cut into the wax with a tiny sapphire, the next hardest jewel to a diamond.

Into his improvements of the phonograph Mr. Edison has put an enormous amount of time and work. He has never lost interest, but has worked on it more or less through all the intervening years up to the present time. Even during recent years he has expended a prodigious amount of energy in improving the reproducer and other parts, spending night after night, and frequently all night, at the laboratory.

Inasmuch as great quantities of phonographs were sold, requiring millions of records, one of the difficulties to be overcome was to make large numbers of duplicates from an original record made by a singer, speaker, or band of musicians.

This difficulty will be perceived when it is stated that the record cut into the wax cylinder is hardly ever greater than one-thousandth of an inch deep, which is less than the thickness of a sheet of tissue paper, and in a single phonograph record there are many millions of sound-waves so recorded.

Through endless experiments of Edison and his working force, and with many ingenious inventions, however, these difficulties were overcome one by one.

It may be added that the phonograph was an invention so absolutely new that when Mr. Edison applied for his original patent, in 1877, the Patent Office could not find that any such attempt had ever before been made to record and reproduce speech or other sounds, and the patent was granted immediately. He has since taken out more than one hundred patents on improvements.

The original patent has long since expired, and many kinds of talking-machines are now made by others also, but they all operate on the identical principle which Edison was the first to discover and put into actual practice.

XIII

A NEW LIGHT IN THE WORLD

In these modern times an incandescent electric lamp is such an every-day affair as to be a familiar object even to a small child. But only a few years ago—a little over thirty—the man who proposed and invented it was derided in the newspapers, and called a madman and a dreamer.

If among Edison's numerous inventions there should be selected one or a class that might be considered the greatest, it seems to be universal opinion that the palm would be awarded to the incandescent lamp and his *complete system* for the distribution of electric light, heat, and power. These inventions as a class, and what has sprung from them, have brought about most wonderful changes in the world.

The year 1877 was a busy one at Edison's laboratory at Menlo Park. He was engaged on the telephone, on acoustic electric transmission, sextuplex telegraphs, duplex telegraphs, miscellaneous carbon articles, and other things. He also commenced experimenting on the electric light.

Besides, as we have seen in the previous chapter, he invented the phonograph. The great interest and excitement caused by the latter invention took up nearly all of his time and attention for many months, and, indeed, up to July, 1878. He then took a vacation and went out to

Wyoming with a party of astronomers to observe an eclipse of the sun and to make a test of his tasimeter.

He was absent about two months, coming home rested and refreshed. Mr. Edison says: "After my return from the trip to observe the eclipse of the sun I went with Professor Barker, professor of physics in the University of Pennsylvania, and Dr. Chandler, professor of chemistry in Columbia College, to see Mr. Wallace, a large manufacturer of brass in Ansonia, Connecticut. Wallace at this time was experimenting on series arc lighting. Just at that time I wanted to take up something new, and Professor Barker suggested that I go to work and see if I could subdivide the electric light so it could be got in small units like gas. This was not a new suggestion, because I had made a number of experiments on electric lighting a year before this. They had been laid aside for the phonograph. I determined to take up the search again and continue it. On my return home I started my usual course of collecting every kind of data about gas; bought all the transactions of the gas engineering societies, etc., all the back volumes of gas journals, etc. Having obtained all the data, and investigated gas-jet distribution in New York by actual observations, I made up my mind that the problem of the subdivision of the electric current could be solved and made commercial."

The problem which Edison had undertaken to solve was a gigantic one. The arc light was then known and in use to a very small extent, but the subdivision of the electric light—as it was then called—had not been accomplished. It had been the dream of scientists and inventors for a long time.

Innumerable trials and experiments had been made in America and Europe for many years, but without success. Although a great number of ingenious lamps had been made by the foremost inventors of the period, they were utterly useless as part of a scheme for a system of electric lighting. In fact, these efforts had been so unsuccessful that many of the leading scientists of the time, even as late as 1879, declared that the subdivision of the light was an impossibility.

The chief trouble was that the early experimenters did not conceive the idea of a *system*, and worked only on a lamp. They all seemed to have the idea that an electric lamp was the main thing and that it should be of low resistance and should be operated on a current of very low voltage, or pressure. They, therefore experimented on lamps using short carbon rods or strips for burners, which required a large quantity of current.

Electric lighting with this kind of lamp was indeed a practical impossibility. The quantity of current required for a large number of them would have been prodigious, giving rise to tremendous problems on account of the heating effects. Besides, the most fatal objection was the cost of copper for conductors, which for a city section of about half a mile square would have cost not less than a hundred million dollars, on account of the enormous quantity of current that would be required.

Mr. Edison realized at the beginning that previous experimenters had failed because they had been following the wrong track. He knew that electric lighting could not be a success unless it could be sold to the public at a reasonable price and pay a profit to those who supplied it. With such lamps as had been proposed, requiring such an enormous outlay for copper, this would have been impossible. Besides, there would not have been enough copper in the world to supply conductors for one large city.

Edison did what he has so often done before and since. He turned about and went in the opposite direction. He reasoned that in order to develop a successful system of electric lighting the cost of conductors must come within very reasonable limits. To insure this, he must invent a lamp of comparatively high resistance, requiring only a small quantity of current, and with a burner having a small radiating surface.

Having the problem clearly in mind, Edison went to work in the fall of 1878 with that enthusiastic energy so characteristic of him. His earliest experiments were made with carbon as the burner for his lamp. In the previous year he had also experimented on this line, beginning with strips of carbon burned in the open air, and then *in vacuo* by means of a hand-worked air-pump. These strips burned only a few minutes. On resuming his work in 1878 he again commenced with carbon, and made a very large number of trials, all *in vacuo*. Not only did he try ordinary strips of carbonized paper, but tissue-paper coated with tar and lampblack was rolled into thin sticks, like knitting-needles, carbonized and raised to the white heat of incandescence *in vacuo*.

He also tried hard carbon, wood carbon, and almost every conceivable variety of paper carbon in like manner. But with the best vacuum that he could then get by means of the ordinary hand-pump the carbons would last at the most only from ten to fifteen minutes in a state of incandescence.

It was evident to Edison that such results as these were not of commercial value. He feared that, after all, carbon was not the ideal substance he had thought it was for an incandescent lamp-burner. The lamp that he had in mind was one which should have a tough, hair-like filament for a light-giving body that could be maintained at a white heat for a thousand hours before breaking.

He therefore turned his line of experiments to wires made of refractory metals, such as platinum and iridium, and their alloys. These metals have very high fusing points, and while they would last longer than the carbon strips, they melted with a slight excess of current after they had been lighted but a short time.

Nevertheless, Edison continued to experiment along this line, making some improvements, until about April, 1879, he made an important discovery which led him to the first step toward the modern incandescent lamp. He discovered that if he introduced a piece of platinum wire into an all-glass globe, completely sealed and highly exhausted of air, and passed a current through the platinum wire while the vacuum was being made the wire would give a light equal to twenty-five candle-power without melting. Previously, the same length of wire would melt in the open air when giving a light equal to four candles.

He thus discovered that the passing of current through the platinum while the vacuum was being obtained would drive out occluded gases (*i.e.*, gases mechanically held in or upon the metal). This was important and soon led to greater results.

Edison and his associates had been working night and day at the Menlo Park laboratory, and now that promising results were ahead their efforts went on with greater vigor than ever. Taking no account of the passage of time, with an utter disregard of meal-times, and with but scanty hours of sleep snatched reluctantly at odd periods, Edison labored on, and the laboratory was kept going without cessation.

Following up the progress he had made, Edison made improvement after improvement, especially in the line of high vacua, and about the beginning of October had so improved his pumps that he could produce a vacuum up to the one-millionth part of an atmosphere. It should be understood that the maintaining of such a high vacuum was only rendered possible by Edison's invention of a one-piece all-glass globe, hermetically sealed during its manufacture into a lamp.

In obtaining this perfection of vacuum apparatus Edison realized that he was drawing nearer to a solution of the problem. For many reasons, however, he was dissatisfied with platino-iridium filaments for burners, and went back to carbon, which from the first he had thought of as an ideal substance for a burner.

His next step proved that he was correct. On October 21, 1879, after many patient trials, he carbonized a piece of cotton sewing-thread bent into a loop or horseshoe form, and had it sealed into a glass globe from which he exhausted the air until a vacuum up to one-millionth of an atmosphere was produced. This lamp, when put on the circuit, lighted up brightly to incandescence and maintained its integrity for over forty hours, and lo! the practical incandescent lamp was born. The impossible, so called, had been attained; subdivision of the electric current was made practicable; the goal had been reached, and one of the greatest inventions of the century was completed.

Edison and his helpers stayed by the lamp during the whole forty hours watching it, some of the men making bets as to how long it would burn. It may well be imagined that there was great jubilation throughout the laboratory during those two days of delight and anxiety.

But now that the principle was established work was renewed with great fervor in making other lamps. A vast number of experiments were made with carbons made of paper, and the manufacture of lamps with these paper carbons was carried on continuously. A great number of these were made and put into actual use.

Edison was not satisfied, however. He wanted something better. He began to carbonize everything that he could lay hands on. In his laboratory note-books are innumerable jottings of the things that were carbonized and tried, such as tissue-paper, soft paper, all kinds of cardboards, drawing paper of all grades, paper saturated with tar, all kinds of threads, fish-line, threads rubbed with tarred lampblack, fine threads plaited together in strands, cotton soaked in boiling tar, lamp-wick, twine, tar and lampblack mixed with a proportion of lime, vulcanized fiber, celluloid, boxwood, cocoanut hair and shell, spruce, hickory, baywood, cedar, and maple shavings, rosewood, punk, cork, bagging, flax, and a host of other things.

He also extended his searches far into the realms of nature in the line of grasses, plants, canes, and similar products, and in these experiments at that time and later he carbonized, made into lamps, and tested no fewer than six thousand different species of vegetable growths.

At this time Edison was investigating everything with a microscope. One day he picked up a palm-leaf fan and examined the long strip of cane binding on its edge. He gave it to one of his assistants, telling him to cut it up into filaments, carbonize them, and put them into lamps.

These proved to be the best thus far obtained, and on further examination Edison decided that he had now found the best material so far tried, and a material entirely suitable for his lamps.

Within a very short time he sent a man off to China and Japan to search for bamboo, with instructions to keep on sending samples until the right one was found. This man did his work well, and among the species of bamboo he sent was one that was found satisfactory. Mr. Edison obtained a quantity of this and arranged with a farmer in Japan to grow it for him and to ship regular supplies. This was done for a number of years, and during that time millions of Edison lamps were regularly made from that particular species of Japanese bamboo.

Mr. Edison did not stop at this, however. He was continually in search of the best, and sent other men out to Cuba, Florida, and all through South America to hunt for something that might be superior to what he was using. Another man was sent on a trip around the world for the same purpose.

Some of these explorers met with striking adventures during their travels, and all of them sent vast quantities of bamboos, palms, and fibrous grasses to the laboratory for examination, but Edison never found any of them better for his purposes than the bamboo from Japan.

In this remarkable exploration of the world for such a material will be found an example of the thoroughness of Edison's methods. He is not satisfied to believe he has the best until he has proved it, and this search for the best bamboo was so thorough that it cost him altogether about one hundred thousand dollars.

In the meantime he was experimenting to manufacture an artificial filament that would be better than bamboo. He finally succeeded in his efforts, and brought out what is known as a "squirted" filament. This was made of a cellulose mixture and pressed out in the form of a thread through dies. This kind of filament has gradually superseded the bamboo in the manufacture of lamps.

We have been obliged to confine ourselves to a very brief outline history of the invention and development of the incandescent lamp. To tell the detailed story of the intense labors of the inventor and his staff of faithful workers would require a volume as large as the present one.

All that could be done in the space at our disposal was to try and give the reader a general idea of the clear thinking, logical reasoning, endless experimenting, hard work, and thoroughness of method of Edison in the creation of a new art.

XIV

MENLO PARK

In the history of the world's progress, Menlo Park, New Jersey, will ever be famous as the birthplace of the carbon transmitter, the phonograph, the incandescent lamp, the commercial dynamo, and the fundamental systems of distributing electric light, heat, and power.

In this list might also be included the electric railway, for while others had previously made some progress in this direction, it was in this historic spot that Edison did his pioneer work that advanced the art to a stage of practicability.

The name of Menlo Park will not have as striking a significance to the younger readers as to their elders whose recollections carry them back to the years between 1876 and 1886. During that period the place became invested with the glamor of romance by reason of the many startling and wonderful inventions coming out of it from time to time.

Edison worked there during these ten years. He had adopted Invention as a profession. As we have seen, he had always had a passion for a laboratory. Thus, from the little cellar at Port Huron, from the scant shelves in a baggage car, from the nooks and corners of dingy telegraph offices, and the grimy little shops in New York and Newark, he had come to the proud ownership of a *real* laboratory where he could wrestle with Nature for her secrets.

Here he could experiment to his heart's content, and invent on a bolder and larger scale than ever before. All the world knows that he did.

Menlo Park was the merest hamlet, located a few miles below Elizabeth. Besides the laboratory buildings, it had only a few houses, the best-looking of which Edison lived in. Two or three of the others were occupied by the families of members of his staff; in the others boarders were taken.

During the ten years that Edison occupied his laboratory there, life in Menlo Park could be summed up in one short word—work. Through the days and through the nights, year in and year out, for the most part, he and his associates labored on unceasingly, snatching only a few hours of sleep here and there when tired nature positively demanded it. Such a scene of concentrated and fruitful activity the world has probably never seen.

The laboratory buildings consisted of the laboratory proper, the library and office, a machine shop, carpenter shop, and some smaller buildings, and, later on, a wooden building, which was used for a short time as an incandescent lamp factory.

Here Edison worked through those busy years, surrounded by a band of chosen assistants, whose individual abilities and never-failing loyalty were of invaluable aid to him in accomplishing the purposes that he had in mind.

As to these associates, we quote Mr. Edison's own words from an autobiographical article in the *Electrical World* of March 5, 1904: "It is interesting to note that in addition to those mentioned above (Charles Batchelor and Francis R.

Upton), I had around me other men who ever since have remained active in the field, such as Messrs. Francis Jahl, William J. Hammer, Martin Force, Ludwig K. Boehm, not forgetting that good friend and co-worker, the late John Kruesi. They found plenty to do in the various developments of the art, and as I now look back I sometimes wonder how we did so much in so short a time."

To this roll of honor may be added the names of a few others: The Carman brothers, Stockton L. Griffin, Dr. A. Haid, John F. Ott (still with Mr. Edison at Orange), John W. Lawson, Edward H. Johnson, Charles L. Clarke, William Holzer, James Hippie, Charles T. Hughes, Samuel D. Mott, Charles T. Mott, E. G. Acheson, Dr. E. L. Nichols, J. H. Vail, W. S. Andrews, and Messrs. Worth, Crosby, Herrick, Hill, Isaacs, Logan, and Swanson.

To these should be added the name of Mr. Samuel Insull, who, in 1881, became Mr. Edison's private secretary, and who for many years afterward managed all his business affairs.

Mr. Insull's position as secretary in the Menlo Park days was not a "soft snap," as his own words will show. He says: "I never attempted to systematize Edison's business life. Edison's whole method of work would upset the system of any office. He was just as likely to be at work in his laboratory at midnight as midday. He cared not for the hours of the day or the days of the week. If he was exhausted he might more likely be asleep in the middle of the day than in the middle of the night, as most of his work in the way of invention was done at night. I used to run his office on as close business methods as my experience admitted, and I would get at him whenever it suited his convenience. Sometimes he would not go over his mail for days at a time, but other times he would go regularly to his office in the morning. At other times my engagements used to be with him to go over his business affairs at Menlo Park at night, if I was occupied in New York during the day. In fact, as a matter of convenience I used more often to get at him at night as it left my days free to transact his affairs, and enabled me, probably at a midnight luncheon, to get a few minutes of his time to look over his correspondence and get his directions as to what I should do in some particular negotiation or matter of finance. While it was a matter of suiting Edison's convenience as to when I should transact business with him, it also suited my own ideas, as it enabled me after getting through my business with him to enjoy the privilege of watching him at his work, and to learn something about the technical side of matters. Whatever knowledge I may have of the electric light and power industry I feel I owe it to the tuition of Edison. He was about the most willing tutor, and I must confess that he had to be a patient one."

It must not be supposed that the hard work of these times made life a burden to the small family of laborers associated with Edison. On the contrary, they were a cheerful, happy lot of men, always ready to brighten up their strenuous life by the enjoyment of anything of a humorous nature that came along.

Often during the long, weary nights of experimenting Edison would call a halt for refreshments, which he had ordered always to be sent in at midnight when night work was in progress. Everything would be dropped, all present would join in the meal, and the last good story or joke would pass around.

Mr. Jehl has written some recollections of this period, in which he says: "Our lunch always ended with a cigar, and I may mention here that although Edison was never fastidious in eating, he always relished a good cigar, and seemed to find in it consolation and solace.... It often happened that while we were enjoying the cigars after our midnight repast, one of the boys would start up a tune on the organ and we would sing together, or one of the others would give a solo. Another of the boys had a voice that sounded like something between the ring of an old tomato-can and a pewter jug. He had one song that he would sing while we roared with laughter. He was also great in imitating the tin-foil phonograph. When Boehm was in good humor he would play his zither now and then, and amuse us by singing pretty German songs. On many of these occasions the laboratory was the rendezvous of jolly and convivial visitors, mostly old friends and acquaintances of Mr. Edison. Some of the office employees would also drop in once in a while, and, as every one present was always welcome to partake of the midnight meal, we all enjoyed these gatherings. After a while, when we were ready to resume work, our visitors would intimate that they were going home to bed, but we fellows could stay up and work, and they would depart, generally singing some song like 'Good-night, Ladies!'... It often happened that when Edison had been working up to three or four o'clock in the morning he would lie down on one of the laboratory tables, and with nothing but a couple of books for a pillow, would fall into a sound sleep. He said it did him more good than being in a soft bed, which spoils a man. Some of the laboratory assistants could be seen now and then sleeping on a table in the early morning hours. If their snoring became objectionable to those still at work, the 'calmer' was applied. This machine consisted of a Babbitt's soap-box without a cover. Upon it was mounted a broad ratchet-wheel with a crank, while into the teeth of the wheel there played a stout, elastic slab of wood. The box would be placed on the table where the snorer was sleeping and the crank turned rapidly. The racket thus produced was something terrible, and the sleeper would jump up as though a typhoon had struck the laboratory. The irrepressible spirit of humor in the old days, although somewhat strenuous at times, caused many a moment of hilarity which seemed to refresh the boys, and enabled them to work with renewed vigor after its manifestation."

The "boys" were ever ready for a joke on one of their number. Mr. Mackenzie, who taught Edison telegraphy, spent a great deal of time at the laboratory. He had a bushy red beard, and was persuaded to give a few hairs to be carbonized and used for filaments in experimental lamps. When the lamps were lighted the boys claimed that their brightness was due to the rich color of the hairs.

The history of the busy years at Menlo Park would make a long story if told in full, but only a hint can be given here of the gradual development of many important inventions. These include the innumerable experiments on the lamp, on different kinds and weights of iron for field magnets and armatures, on magnetism, on windings and connections for field magnets and armatures, on distribution circuits, control, and regulation, and so on through a long list.

All these things were new. There was nothing in the books to serve as a guide in solving these new problems, but Edison patiently worked them out, one by one, until a complete system was the result of his labors.

Menlo Park was historic in one other particular. It was the very first place in the world to see incandescent electric lighting from a central station.

The newspapers had been so full of the wonderful invention that there was a great demand to see the new light. Edison decided to give a public exhibition, and for this purpose put up over four hundred lights in the streets and houses of Menlo Park, all connected to underground conductors which ran to the dynamos in one of the shop buildings.

On New Year's Eve, 1879, the Pennsylvania Railroad ran special trains, and over three thousand people availed themselves of the opportunity to witness the demonstration. It was a great success, and gave rise to a wide public interest.

Edison's laboratory at Menlo Park had never suffered for lack of visitors, but now it became a center of attraction for scientific and business men from all parts of the world. Pages of this book could be filled with the names of well-known visitors at this period, but it would be of no practical use to give them; besides we must now pass on to the time when the light was introduced to the world.

XV

BEGINNING THE ELECTRIC LIGHT BUSINESS

The close of the last two chapters found us attending the birth of an art that was then absolutely and entirely new—the art of electric lighting by incandescent lamps. It will now be interesting to take a brief glance at the way in which it was introduced to the world.

Edison invented not only a lamp and a dynamo, but a complete *system* of distributing electric light, heat, and power from central stations. This included a properly devised network of conductors fed with electricity from several directions and capable of being tapped to supply current to each building; a lamp that would be cheap, lasting, take little current, be easy to handle, and each to be independent of every other lamp; means for measuring electricity by meter; means for regulating the current so that every lamp, whether near to or far away from the station, would give an equal light; the designing of new and efficient dynamos, with means for connecting and disconnecting and for regulating and equalizing their loads; the providing of devices that would prevent fires from excessive current, and the providing of switches, lamp-holders, fixtures, and the like.

This was a large program to fill, for it was all new, and there was nothing in the world from which to draw ideas, but Edison carried out his scheme in full, and much more besides. By the end of 1880 he was ready to launch his electric light system for commercial use, and the Edison Electric Light Company, that had been organized for the purpose, rented a mansion at No. 65 Fifth Avenue, New York, to be used for offices. Edison now moved some of his Menlo Park staff into that city to pursue the work.

Right at the very beginning a most serious difficulty was met with. None of the appliances necessary for use in the lighting system could be purchased anywhere in the world.

They were all new and novel—dynamos, switchboards, regulators, pressure and current indicators, incandescent lamps, sockets, small switches, meters, fixtures, underground conductors, junction boxes, service boxes, manhole boxes, connectors, and even specially made wire. Not one of these things was in existence; and no outsider knew enough about such devices to make them on order, except the wire.

Edison himself solved the difficulty by raising some money and establishing several manufacturing shops in which these articles could be made. The first of all was a small factory at Menlo Park to make the lamps, Mr. Upton taking charge of that branch.

For making the dynamos he secured a large works on Goerck Street, New York, and gave its management to Mr. Batchelor. For the underground conductors and their parts a building on Washington Street was rented and the work done under the superintendence of Mr. Kruesi. In still another factory building there were made the smaller appliances, such as sockets, switches, fixtures, meters, safety fuses and other details. This latter plant was at first owned by Mr. Sigmund Bergmann, who had worked with Edison on telephones and phonographs, but later Mr. Edison and E. H. Johnson became partners.

Still another difficulty presented itself. There were no men who knew how to do wiring for electric lights, except those who had been with Edison at Menlo Park. This problem was solved by opening a night-school at No. 65 Fifth Avenue in which a large number of men were educated and trained for the work by Edison's associates. Many of these men have since become very prominent in electrical circles.

Thus, in planning these matters, and in guiding the operations in these four shops in New York, and with all the work he was doing on new experiments and inventions there and at Menlo Park, and in making preparations for the first central station in New York City, Edison was a prodigiously busy man. He worked incessantly, and it is safe to say that he did not average more than four hours' sleep a day.

He was the center and the guiding spirit of those intensely busy times. The aid of his faithful associates was invaluable in the building up of the business, but he was the great central storehouse of ideas, and it is owing to his undaunted courage, energy, perseverance, knowledge and foresight, that the foundations of so great an art have been so well laid.

As has been well said by Major S. B. Eaton, who was president and general manager of the Edison Electric Light Company in its earliest years: "In looking back on those days and scrutinizing them through the years, I am impressed by the greatness, the solitary greatness, I may say, of Mr. Edison. We all felt then that we were of importance, and that our contribution of effort and zeal was vital. I can see now, however, that the best of us was nothing but the fly on the wheel. Suppose anything had happened to Edison? All would have been chaos and ruin. To him, therefore, be the glory, if not the profit."

Early in 1881 comparatively few people had seen the incandescent light. In order to make the public familiar with it, the Edison company equipped its office building with fixtures and lamps, the latter being lighted by current from a dynamo in the cellar. In the evenings the house was thrown open to visitors until ten or eleven o'clock. Thousands of people flocked to see the new light, which in those days was regarded as wonderful and mysterious, for while the lamps gave a soft, steady illumination, there was no open flame, practically no heat, no danger of fire, and no vitiation of air. For the most part of four years the writer spent his evenings receiving these visitors if no important business was in progress at the moment.

Mr. Edison and his shops had scarcely time to get well on their feet before a rush of business set in. How this business rapidly developed and grew until it became of very great magnitude is a matter of history, which we shall not attempt to relate here.

Some idea of this wonderful development, as it has gone on through the years that have passed since 1880, may be formed when it is stated that at this time there are more than one hundred millions of incandescent lamps in daily use in the United States alone. Every one of these lamps and the fundamental principles upon which they are operated rest upon the foundations which Edison laid so well.

One of Mr. Edison's interesting stories of the early days relates to the making of the lamps. He says:

"When we first started the electric light we had to have a factory for manufacturing lamps. As the Edison light company did not seem disposed to go into manufacturing, we started a small lamp factory at Menlo Park with what money I could raise from my other inventions and royalties and some assistance. The lamps at that time were costing about one dollar and twenty-five cents each to make, so I said to the company: 'If you will give me a contract during the life of the patents I will make all the lamps required by the company and deliver them for forty cents.' The company jumped at the chance of this offer, and a contract was drawn up. We then bought at a receiver's sale at Harrison, New Jersey, a very large brick factory building which had been used as an oil-cloth works. We got it at a great bargain, and only paid a small sum down, and the balance on mortgage. We moved the lamp works from Menlo Park to Harrison. The first year the lamps cost us about one dollar and ten cents each. We sold them for forty cents; but there were only about twenty or thirty thousand of them. The next year they cost us about seventy cents, and we sold them for forty. There were a good many, and we lost more money the second year than the first. The third year I succeeded in getting up machinery and in changing the processes, until it got down so that they cost somewhere around fifty cents. I still sold them for forty cents, and lost more money that year than any other, because the sales were increasing rapidly. The fourth year I got it down to thirty-seven cents, and I made all the money in one year that I had lost previously. I finally got it down to twenty-two cents, and sold them for forty cents; and they were made by the million. Whereupon the Wall Street people thought it was a very lucrative business, so they concluded they would like to have it, and bought us out.

"When we formed the works at Harrison we divided the interests into one hundred shares or parts at one hundred dollars par. One of the boys was hard up after a time, and sold two shares to Bob Cutting. Up to that time we had never paid anything, but we got around to the point where the board declared a dividend every Saturday night. We had never declared a dividend when Cutting bought his shares, and after getting his dividends for three weeks in succession he called up on the telephone and wanted to know what kind of a concern this was that paid a weekly dividend. The works sold for $1,085,000."

We have been obliged to confine ourselves to a very brief and general description of the beginnings of the art of electric lighting, but this chapter would not be complete without reference to Edison's design and construction of the greatest dynamo that had ever been made up to that time.

The earliest dynamos he made would furnish current only for sixty lamps of sixteen candle-power each. These machines were belted up to an engine or countershaft. He realized that much larger dynamos would be needed for central stations, and in 1880 constructed one in Menlo Park, but it was not entirely successful.

In the spring of 1881, however, he designed a still larger one, to be connected direct to its own engine and operated without belting. Its capacity was to be twelve hundred lamps, instead of sixty.

At that time such a project was not dreamed of outside the Edison laboratory, and once more he was the subject of much ridicule and criticism by those who were considered as experts. They said the thing was impossible and absolutely impracticable.

Such opinions, however, have never caused a moment's hesitation to Edison when he has made up his mind that a thing can be done. He calmly went ahead with his plans, and although he found many difficulties, he overcame them all. He worked the shops night and day, until he had built this great machine and operated it successfully.

The dynamo was finished in the summer of 1881. At that time there was in progress an international Electrical Exposition in Paris, at which Edison was exhibiting his system of electric lighting. He had promised to send this great dynamo over to Paris.

When the dynamo was finished and tested there were only four hours to take it and the engine apart and get all the parts on board the steamer. Edison had foreseen all this, and had arranged to have sixty men get to work all at once to take it apart. Each man had written instructions just what to do, and when the machine was stopped every man did his own particular work and the job was quickly accomplished.

Arrangements had been made with the police for rapid passage through the streets from the shops to the steamship. The trucks made quick time of it, being preceded by a wagon with a clanging bell. Street traffic was held up for them, just as it is for engines and hose-carts going to a fire. The dynamo and engine got safely down to the dock without delay and were loaded on the steamer an hour before she sailed.

This dynamo and engine weighed twenty-seven tons, and was then, and for a long time after, the eighth wonder of the scientific world. Its arrival and installation in Paris were eagerly watched by the most famous scientists and electricians in Europe.

XVI

THE FIRST EDISON CENTRAL STATION

From the beginning of his experiments on the electric light Edison had one idea ever in mind, and that was to develop a system of lighting cities from central stations. His plan was to supply electric light and power in much the same way that gas is furnished.

He never forsook this idea for a moment. Indeed, it formed the basis of all his plans, although the scientific experts of the time predicted utter failure. While the experiments were going on at Menlo Park he had Mr. Upton and others at work making calculations and plans for city systems.

Soon after he had invented the incandescent lamp he began to take definite steps toward preparing for the first central station in the city of New York. After some consideration, he decided upon the district included between Wall, Nassau, Spruce and Ferry Streets, Peck Slip and the East River, covering nearly a square mile in extent.

He sent into this district a number of men, who visited every building, counted every gas-jet and found out how many hours per day or night they were burned.

These men also ascertained the number of business houses using power and how much they consumed. All this information was marked in colored inks on large maps, so that Edison could study the question with all the details before him.

All this work had taken several months, but, with this information to guide him, the main conductors to be laid in the streets of this district were figured, block by block, and the results were marked upon the maps. It was found, however, that the quantity of copper required for these conductors would be exceedingly large and costly, and, if ever, Edison was somewhat dismayed.

This difficulty only spurred him on to still greater effort. Before long he solved the problem by inventing the "feeder and main" system, for which he signed an application for patent on August 4, 1880.

By this invention he saved seven-eighths of the amount of copper previously required. So the main conductors were figured again, at only one-eighth the size they were before, and the results were marked upon enormous new maps which were now prepared for the actual installation.

It should be remembered that from the very start Edison had determined that his conductors should be placed underground. He knew that this was the only method for permanent and satisfactory service to the public.

Our young readers can scarcely imagine the condition of New York streets at that time. They were filled with lines of ugly wooden poles carrying great masses of telegraph, telephone, stock ticker, burglar alarm and other wires, in all conditions of sag and decay. The introduction of the arc-lamp added another series of wires which with their high potentials carried a menace to life. Edison was the first to put conductors underground, and the wisdom of so doing became so clear that a few years later laws were made compelling others to do likewise.

But to return to our story. Just before Christmas in 1880 the Edison Electric Illuminating Company of New York was organized, and a license was issued to it for the use of the Edison patents on Manhattan Island.

The work for the new station now commenced in real earnest. A double building at 255 and 257 Pearl Street was purchased, and the inside of one half was taken out and a strong steel structure was erected inside the walls.

Work on the maps and plans for the underground network of conductors was continued at Menlo Park. Mr. Edison started his factories for making dynamos, lamps, underground conductors, sockets, switches, meters, and other details. Thus, the wheels of industry were humming merrily in preparation for the installation of the system. Every detail received Edison's personal care and consideration. He had plenty of competent men, but he deemed nothing too small or insignificant for his attention in this important undertaking.

In the fall of 1881 the laying of the underground conductors was begun and pushed forward with frantic energy. Here again Edison left nothing to chance. Although he had a thousand things to occupy his mind he also superintended this work. He did not stand around and give orders, but worked with the men in the trenches day and night helping to lay tubes, filling up junction boxes, and taking part in all the infinite detail.

He would work till he felt the need of a little rest. Then he would go off to the station building in Pearl Street, throw an overcoat on a pile of iron tubes, lie down and sleep a few hours, rising to resume work with the first gang.

It is worth pausing just a moment to glance at this man taking a fitful rest on a pile of iron pipe in a dingy building. His name is on the tip of the world's tongue.

Distinguished scientists from every part of Europe seek him eagerly. He has just been decorated and awarded high honors by the French government. He is the inventor of wonderful new apparatus and the exploiter of novel and successful arts. The magic of his achievements and the rumors of what is being done have caused a wild drop in gas securities and a sensational rise in his own electric-light stock from one hundred dollars to thirty-five hundred a share. Yet these things do not at all affect his slumber or his democratic simplicity, for in that, as in everything else, he is attending strictly to business, "doing the thing that is next to him."

The laying of the underground conductors was interrupted by frost in the winter of 1881, but in the following spring the work was renewed with great energy until there had been laid over eighty thousand feet. In the mean time the buildings of the district were being wired for lamps, and the machine-works had been busy on the building of three of the "Jumbo" dynamos for the station. These were larger than the great dynamo that had been sent to Paris.

These three dynamos were installed in the station, and the other parts of the system were completed. A bank of one thousand lamps was placed in one of the buildings; and in the summer a whole month was spent in making tests of the working of the system, using this bank of lamps instead of sending current out to customers' premises. Edison and his assistants made the station their home during this busy month. They even slept there on cots that he had sent to the station for this purpose.

The system tested out satisfactorily, and finally, on September 4, 1882, at three o'clock in the afternoon, the station was started by sending out current from one of the big dynamos through the conductors laid in the streets, and electric light was supplied for the first time to a number of customers in the district.

The station was now started and everything went well. New customers were added daily, and very soon it became necessary to supply more current. This called for the operation of two dynamos at one time. As this involved new problems, Edison chose a Sunday to try it, when business places would be closed. We will let him tell the story. He says: "My heart was in my mouth at first, but everything worked all right.... Then we started another engine and threw the dynamos in parallel. Of all the circuses since Adam was born, we had the worst then! One engine would stop, and the other would run up to about a thousand revolutions, and then they would see-saw. The trouble was with the governors. When the circus commenced the gang that was standing around ran out precipitately, and I guess some of them kept running for a block or two. I grabbed the throttle of one engine, and E. H. Johnson, who was the only one present to keep his wits, caught hold of the other, and we shut them off."

One of the gang that ran, but, in this case, only to the end of the room, afterward said: "At the time it was a terrifying experience, as I didn't know what was going to happen. The engines and dynamos made a horrible racket, from loud and deep groans to a hideous shriek, and the place seemed to be filled with sparks and flames of all colors. It was as if the gates of the infernal regions had been suddenly opened."

Edison attacked this problem in his strenuous way. Although it was Sunday, he sent out and gathered his men and opened the machine-works to make new appliances to overcome this trouble.

Space will not permit of telling all the methods he applied until the difficulty was entirely conquered. It was only a short time, however, before he was able to operate two or any number of dynamos all together as one, in parallel, without the least trouble.

This early station grew and prospered, and continued in successful operation for more than seven years, until January 2, 1890, when it was partially destroyed by fire. This occurrence caused a short interruption of service, but in a few days current was again supplied to customers as before, and the service has never since ceased.

Increasing demands for service soon afterward led to the construction of other stations on Manhattan Island, until at the present time the New York Edison Company (the successor to the Edison Electric Illuminating Company of New York) is operating over forty stations and sub-stations. These supply current for about 800,000 customers, wired for 17,000,000 incandescent lamps and for about 1,300,000 horse-power in electric motors.

The early success of the first central station in New York led to the formation of new companies in other cities, and the installation of many similar plants. The business has grown by leaps and bounds, until at the present time there are many thousands of central stations spread all over the United States, furnishing electric light, heat, and power, chiefly by use of the principles elaborated so many years ago by Mr. Edison.

We ought to mention that this tremendous growth has also been largely due to another invention made by him in 1882, called the "three-wire system." Its value consists in the fact that it allowed a further saving of sixty-two and one-half per cent, of copper required for conductors. This invention is in universal use all over the world.

It may be mentioned here that at the opening ceremonies of the Electrical Exposition in New York, on October 11, 1911, the leading producers and consumers of copper presented Mr. Edison with an inscribed cubic foot of that metal in recognition of the stimulus of his inventions to the industry. The inscription shows that the yearly output of copper was 377,644,000 pounds at the time of Edison's first invention in 1868, and in October, 1911, the yearly output had increased to 1,910,608,000 pounds.

XVII

EDISON'S ELECTRIC RAILWAY

It is quite likely that many of our young readers have never seen a horse-car. This is not strange, for in a little over twenty years the victorious trolley has displaced the old-time street-cars drawn by one or two horses. Indeed, a horse-car is quite a curiosity in these modern days, for such vehicles have almost entirely disappeared from the streets.

The first horse railroad in the United States was completed in 1827, and it was only seven years afterward that a small model of a circular electric railroad was made and exhibited by Thomas Davenport, of Brandon, Vermont. Other inventors also worked on electric railways later on, but they did not make much progress, because in their day there were no dynamos, and they had to use primary batteries to obtain current. This method of generating current was far too cumbersome and expensive for general use.

In 1879, after dynamos had become known, the firm of Siemens exhibited at the Berlin Exhibition a road about one-third of a mile in length, over which an electric locomotive hauled three small cars at a speed of about eight miles an hour.

This was just before Edison had developed the efficient commercial dynamo with low-resistance armature and high-resistance field, which made it possible to generate and use electric power cheaply. Thus we see that Edison was not the first to form the broad idea of a electric railway, but his dynamo and systems of distribution and regulation of current first made the idea commercially practicable.

When Edison made his trip to Wyoming with the astronomers in 1878 he noticed that the farmers had to make long hauls of their grain to the railroads or markets. He then conceived the idea of building light electric railways to perform this service.

As we have already noted, he started on his electric-light experiments, including the dynamo, when he returned from the West. He had not forgotten his scheme for an electric railway, however, for, early in 1880, after the tremendous rush on the invention of the incandescent lamp had begun to subside, he commenced the construction of a stretch of track at Menlo Park, and at the same time began to build an electric locomotive to operate over it.

The locomotive was an ordinary flat dump-car on a four-wheeled iron truck. Upon this was mounted one of his dynamos, used as a motor. It had a capacity of about twelve horse-power. Electric current was generated by two dynamos in the machine-shop, and carried to the rails by underground conductors.

The track was about a third of a mile in length, the rails being of light weight and spiked to ties laid on the ground. In this short line there were some steep grades and short curves. The locomotive pulled three cars; one a flat freight-car; one an open awning-car, and one box-car, facetiously called the "Pullman," with which Edison illustrated a system of electromagnetic braking.

THE EDISON ELECTRIC RAILWAY AT MENLO
PARK—1880

On May 13, 1880, this road went into operation. All the laboratory "boys" made holiday and scrambled aboard for a trip. Things went well for a while, but presently a weakness developed and it became necessary to return the locomotive to the shop to make changes in the mechanism. And so it was for a short time afterward. Imperfections of one kind and another were disclosed as the road was operated, but Edison was equal to the occasion and overcame them, one by one. Before long he had his locomotive running regularly, hauling the three cars with freight and passengers back and forth over the full length of the track. Incidentally, the writer remembers enjoying a ride over the road one summer afternoon.

The details of the various improvements made during these months are too many and too technical to be given here. It is a fact, however, that at this time Edison was doing some heavy electric railway engineering, each improvement representing a step which advanced the art toward the perfection it has reached in these modern days.

The newspapers and technical journals lost no time in publishing accounts of this electric railroad, and once again Menlo Park received great numbers of visitors, including many railroad men, who came to see and test this new method of locomotion.

Of course, in operating this early road there were a few mishaps, fortunately none of them of a serious nature. In the correspondence of the late Grosvenor P. Lowry, a friend and legal adviser of Mr. Edison, is a letter dated June 5, 1880, giving an account of one experience. The letter reads as follows: "Goddard and I have spent a part of the day at Menlo, and all is glorious. I have ridden at forty miles an hour on Mr. Edison's electric railway—and we ran off the track. I protested at the rate of speed over the sharp curves, designed to show the power of the engine, but Edison said they had done it often. Finally, when the last trip was to be taken, I said I did not like it, but would go along. The train jumped the track on a short curve, throwing Kruesi, who was driving the engine, with his face down in the dirt, and another man in a comical somersault through some underbrush. Edison was off in a minute, jumping and laughing, and declaring it a most beautiful accident. Kruesi got up, his face bleeding, and a good deal shaken; and I shall never forget the expression of voice and face in which he said, with some foreign accent: 'Oh yes! pairfeckly safe.' Fortunately no other hurts were suffered, and in a few minutes we had the train on the track and running again."

This first electric railway was continued in operation right along through 1881. In the fall of that year Edison was requested by the late Mr. Henry Villard to build a longer road at Menlo Park, equipped with more powerful locomotives, to demonstrate the feasibility of putting electric railroads in the Western wheat country.

Work was commenced at once, and early in 1882 the road and its equipment were finished. It was three miles long, and had sidings, turn-tables, freight platform and car-house. It was much more complete and substantial than the first railroad. There were two locomotives, one for freight and the other for passenger service.

The passenger locomotive was very speedy and hauled as many as ninety persons at a time. Many thousands of passengers traveled over the road during 1882. The freight locomotive was not so speedy, but could pull heavy trains at a good speed. Taken altogether, this early electric railway made a great advance toward modern practice as its exists to-day.

There are many interesting stories of the railway period at Menlo Park. One of them, as told by the late Charles T. Hughes, who worked with Edison on the experimental roads, is as follows: "Mr. Villard sent J. C. Henderson, one of his mechanical engineers, to see the road when it was in operation, and we went down one day—Edison, Henderson, and I—and went on the locomotive. Edison ran it, and just after we started there was a trestle sixty feet long and seven feet deep, and Edison put on all the power. When we went over it we must have been going forty miles an hour, and I could see the perspiration come out on Henderson. After we got over the trestle and started on down the track Henderson said: 'When we go back I will walk. If there is any more of that kind of running I won't be in it myself.'"

The young reader, who is now living in an age in which the electric railway is regarded as a matter of course, will find it difficult to comprehend that there should ever have been any doubt on the part of engineering experts as to the practicability of electric railroads. But in the days of which we are writing such was the case, as the following remarks of Mr. Edison will show: "At one time Mr. Villard got the idea that he would run the mountain division of the Northern Pacific Railroad by electricity. He asked me if it could be done. I said: 'Certainly; it is too easy for me to undertake; let some one else do it.' He said: 'I want you to tackle the problem,' and he insisted on it. So I got up a scheme of a third rail and shoe and erected it in my yard here in Orange. When I got it all ready he had all his division engineers come on to New York, and they came over here. I showed them my plans, and the unanimous decision of the engineers was that it was absolutely and utterly impracticable. That system is on the New York Central now, and was also used on the New Haven road in its first work with electricity."

Mr. Edison knew at the time that these engineers were wrong. They were prejudiced and lacking in foresight, and had no faith in electric railroading. Indeed, these particular engineers were not by any means the only persons who could see no future for electric methods of transportation. Their doubts were shared by capitalists and others, and it was not until several years afterward that the business of electrifying street railroads was commenced in real earnest.

In the mean time, however, Edison's faith did not waver, and he continued his work on electric railways, making innumerable experiments and taking out a great many patents, including a far-sighted one covering a sliding contact in a slot. This principle and many of those covered by his earlier work are in use to-day on the street railways in large cities.

The early railroad at Menlo Park has gone to ruin and decay, but the crude locomotive built by Edison has become the property of the Pratt Institute, of Brooklyn, New York, to whose students it is a constant example and incentive.

Down to the present moment Edison has kept up an active interest in transportation problems. His latest work has been in the line of operating street-cars with his improved storage battery. During the time that this book has been in course of preparation he has given a great deal of time to this question.

Some years ago there were a number of street-cars in various cities operated by storage batteries of a class entirely different from the battery invented by Edison. We refer to storage batteries containing lead and sulphuric acid. These were found to be so costly to operate and maintain that their use was abandoned.

Mr. Edison's new nickel and iron storage battery with alkaline solution has been found by practical use to be entirely satisfactory for operating street-cars, not only at a low cost, but also with ease of operation and at a trifling expense for maintenance. Of course there have been many problems, but he has surmounted the principal difficulties, and there are now quite a number of street-cars operated by his storage battery in various cities. These cars are earning profits and their number is steadily increasing.

XVIII

GRINDING MOUNTAINS TO DUST

On walking along the sea-shore the reader may have noticed occasional streaks or patches of bluish-black sand, somewhat like gunpowder in appearance. It is carried up from the bed of the sea and deposited by the waves on the shore to a greater or lesser extent on many beaches.

If a magnet be brought near to this "black sand" the particles will be immediately attracted to it, just as iron filings would be in such a case. As a matter of fact, these particles of black sand are grains of finely divided magnetic iron in a very pure state.

Now, if we should take a piece of magnetic iron ore in the form of a rock and grind it to powder the particles of iron could be separated from the ground-up mass by drawing them out with a magnet, just as they could be drawn out of a heap of seashore sand. If all the grains of iron were thus separated and put together, or concentrated, they would be called concentrates.

During the last century a great many experimenters besides Edison attempted to perfect various cheap methods of magnetically separating iron ores, but until he took up the work on a large scale no one seems to have realized the real meaning of the tremendous problems involved.

The beginning of this work on the part of Edison was his invention in 1880 of a peculiar form of magnetic separator. It consisted of a suspended V-shaped hopper with an adjustable slit along the pointed end. A long electromagnet was placed, edgewise, a little below the hopper, and a bin with a dividing partition in the center was placed on the floor below.

Crushed ore, or sand, was placed in the hopper. If there was no magnetism this fine material would flow down in a straight line past the magnet and fill the bin on one side of the partition. If, however, the magnet were active the particles of iron would be attracted out of the line of the falling material, but their weight would carry them beyond the magnet and they would fall to the other side of the partition. Thus, the material would be separated, the grains of iron going to one side and the grains of rock or sand to the other side.

This separator, as afterward modified, was the basis of a colossal enterprise conducted by Mr. Edison, as we shall presently relate. But first let us glance at an early experiment on the Atlantic seashore in 1881, as mentioned by him. He says:

"Some years ago I heard one day that down at Quogue, Long Island, there were immense deposits of black magnetic sand. This would be very valuable if the iron could be separated from the sand. So I went down to Quogue with one of my assistants and saw there for miles large beds of black sand on the beach in layers from one to six inches thick—hundreds of thousands of tons. My first thought was that it would be a very easy matter to concentrate this, and I found I could sell the stuff at a good price. I put up a small magnetic separating plant, but just as I got it started a tremendous storm came up, and every bit of that black sand went out to sea. During the twenty-eight years that have intervened it has never come back."

In the same year a similar separating plant was put up and worked on the Rhode Island shore by the writer under Mr. Edison's direction. More than one thousand tons of concentrated iron ore of fine quality were separated from sea-shore sand and sold. It was found, however, that it could not be successfully used on account of being so finely divided. Had this occurred a few years later, when Edison invented a system of putting this fine ore into briquettes, that part of the story might have been different.

Magnetic separation of ores was allowed to rest for many years after this, so far as Edison was concerned. He was intensely busy on the electric light, electric railway, and other similar problems until 1888, and then undertook the perfecting and manufacturing of his improved phonograph, and other matters. Somewhere about 1890, however, he again took up the subject of ore-separation.

For some years previous to that time the Eastern iron-mills had been suffering because of the scarcity of low-priced high-grade ores. If low-grade ores could be crushed and the iron therein concentrated and sold at a reasonable price the furnaces would be benefited. Edison decided, after mature deliberation, that if these low-grade ores were magnetically separated on a colossal scale at a low cost the furnace-men could be supplied with the much-desired high quality of iron ore at a price which would be practicable.

He appreciated the fact that it was a serious and gigantic problem, but was fully satisfied that he could solve it. He first planned a great magnetic survey of the East, with the object of locating large bodies of magnetic iron ore. This survey was the greatest and most comprehensive of the kind ever made. With a peculiarly sensitive magnetic needle to indicate the presence of magnetic ore in the earth, he sent out men who made a survey of twenty-five miles across country, all the way from lower Canada to North Carolina.

Edison says: "The amount of ore disclosed by this survey was simply fabulous. How much so may be judged from the fact that in the three thousand acres immediately surrounding the mills that I afterward established at Edison, New Jersey, there were over two hundred million tons of low-grade ore. I also secured sixteen thousand acres in which the deposit was proportionately as large. These few acres alone contained sufficient ore to supply the whole United States iron trade, including exports, for seventy years."

Given a mountain of rock containing only one-fifth to one-fourth magnetic iron, the broad problem confronting Edison resolved itself into three distinct parts—first, to tear down the mountain bodily and grind it to powder; second, to extract from this powder the particles of iron mingled in its mass; and third, to accomplish these results at a cost sufficiently low to give the product a commercial value.

From the start Edison realized that in order to carry out this program there would have to be automatic and continuous treatment of the material, and that he would have to make the fullest possible use of natural forces, such as gravity and momentum. The carrying out of these principles and ideas gave rise to some of the most brilliant engineering work that has ever been done by Edison. During this period he also made many important inventions, of which several will now be mentioned.

As he proposed to treat enormous masses of material, one of the chief things to be done was to provide for breaking the rock and crushing it to powder rapidly and cheaply. After some experimenting, he found there was no machinery to be bought that would do the work as it must be done. He was therefore compelled to invent a series of machines for the purpose.

The first of these was an invention quite characteristic of Edison's daring and boldness. It embraced a gigantic piece of mechanism, called the "Giant Rolls," which was designed to break up pieces of rock that might be as large as an ordinary upright piano, and weighing as much as eight tons.

A pair of iron cylinders five feet long and six feet in diameter, covered with steel knobs, were set fifteen inches apart in a massive frame. The rolls weighed about seventy tons. By means of a steam engine these rolls were revolved in opposite directions until they attained a peripheral speed of about a mile a minute. Then the rocks were dumped into a hopper which guided them between the rolls, and in a few seconds, with a thunderous noise, they were reduced to pieces about the size of a man's head. The belts were released by means of slipping friction clutches when the load was thrown on the rolls, the breaking of the rocks being accomplished by momentum and kinetic energy.

The broken rock then passed through similar rolls of a lesser size, by means of which it was reduced to much smaller pieces. These in their turn passed through a series of other machines in which they were crushed to fine powder. Here again Edison made another remarkable invention, called the "Three-High Rolls," for reducing the rock to fine powder. The best crushers he had been able to buy had an efficiency of only eighteen per cent, and a loss by friction of eighty-two per cent. By his invention he reversed these figures and obtained a working efficiency of eighty-four per cent, and reduced the loss to sixteen per cent.

The problems of drying and screening the broken and crushed material were also solved most ingeniously by Edison's inventive skill and engineering ability, and always with the idea and purpose in mind of accomplishing these results by availing himself to the utmost of one of the great forces of Nature—gravity.

The great extent of the concentrating works may be imagined when we state that two hundred and fifty tons of material per hour could be treated. Altogether, there were about four hundred and eighty immense magnetic separators in the plant, through which this crushed rock passed after going through the numerous crushing, drying, and screening processes.

EDISON AT THE OFFICE DOOR OF THE ORE-CONCENTRATING PLANT AT EDISON, NEW JERSEY, IN THE 'NINETIES

If it had been necessary to transfer this tremendous quantity of material from place to place by hand the cost would have been too great. Edison, therefore, designed an original and ingenious system of mechanical belt conveyors that would automatically receive and discharge their loads at appointed places in the works, covering about a mile in transit. They went up and down, winding in and out, turning corners, delivering material from one bin to another, making a number of loops in the drying-oven, filling up bins, and passing on to the next one when full. In fact, these conveyors in automatic action seemed to play their part with human intelligence.

We have been able to take only a passing glance at the great results achieved by Edison in his nine years' work on this remarkable plant—a work deserving of most serious study. The story would be incomplete, however, if we did not mention his labors on putting the fine ore in the form of solid briquettes.

When the separated iron was first put on the market it was found that it could not be used in that form in the furnaces. Edison was therefore obliged to devise some other means to make it available. After a long series of experiments he found a way of putting it into the form of small, solid briquettes. These answered the purpose exactly.

This called for a line of new machinery, which he had to invent to carry out the plan. When this was completed, the great rocks went in at one end of the works and a stream of briquettes poured out of the other end, being made by each briquetting machine at the rate of sixty per minute.

Thus, with never-failing persistence, infinite patience, intense thought and hard work, Edison met and conquered, one by one, the difficulties that had confronted him. Furnace trials of his briquettes proved that they were even better than had been anticipated. He had received some large orders for them and was shipping them regularly. Everything was bright and promising, when there came a fatal blow.

The discovery of rich Bessemer ore in the Mesaba range of mountains in Minnesota a few years before had been followed by the opening of the mines there about this time. As this rich ore could be sold for three dollars and fifty cents per ton, as against six dollars and fifty cents per ton for Edison's briquettes, his great enterprise must be abandoned at the very moment of success.

It was a sad blow to Edison's hopes. He had spent nine years of hard work and about two millions of his own money in the great work that had thus been brought to nought through no fault of his. The project had lain close to his heart and ambition, indeed he had put aside almost all other work and inventions for a while.

For five of the nine years he had lived and worked steadily at Edison (the name of the place where the works were located), leaving there only on Saturday night to spend Sunday at his home in Orange, and returning to the plant by an early train on Monday morning. Life at Edison was of the simple kind—work, meals, and a few hours' sleep day by day, but Mr. Edison often says he never felt better than he did during those five years.

After careful investigations and calculations it was decided to close the plant. Mr. W. S. Mallory, his close associate during those years of the concentrating work, says: "The plant was heavily in debt, and, as Mr. Edison and I rode on the train to Orange, plans were discussed as to how to make enough money to pay off the debt. Mr. Edison stated most positively that no company with which he had been personally actively connected had ever failed to pay its debts, and he did not propose to have the concentrating company any exception.

"We figured carefully over the probabilities of financial returns from the phonograph works and other enterprises, and, after discussing many plans, it was finally decided that we would apply the knowledge we had gained in the concentrating plant to building a plant for manufacturing Portland cement, and that Mr. Edison would devote his attention to the developing of a storage battery which did not use lead and sulphuric acid.

"He started in with the maximum amount of enthusiasm and ambition, and in the course of about three years we succeeded in paying off the indebtedness of the concentrating works.

"As to the state of Mr. Edison's mind when the final decision was reached to close down, if he was specially disappointed there was nothing in his manner to indicate it, his every thought being for the future."

In this attitude we find a true revelation of one conspicuous trait in Mr. Edison. No one ever cried less over spilled milk than he. He had spent a fortune and had devoted nine years of his life to the most intense thought and labor in the creation and development of this vast enterprise. He had made many remarkable inventions and had achieved a very great success, only to see the splendid results swept away in a moment. He did not sit down and bewail his lot, but with true philosophy and greatness of mind applied himself with characteristic energy to new work through which he might be able to open up a more promising future.

XIX

EDISON MAKES PORTLAND CEMENT

Long before Edison ever thought of going into the manufacture of cement he had very pronounced opinions of its value for building purposes. More than twenty-five years ago, during a discussion on ancient buildings, he remarked: "Wood will rot, stone will chip and crumble, bricks disintegrate, but a cement and iron structure is apparently indestructible. Look at some of the old Roman baths. They are as solid as when they were built."

With such convictions, and the vast fund of practical knowledge and experience he had gained at Edison in the crushing and handling of enormous masses of finely divided material, it is not surprising that he should have decided to engage in the manufacture of cement.

He was fully aware of the fact that he was proposing to "butt into" an old-established industry, in which the principal manufacturers were concerns which had been in business for a long time. He knew there were great problems to be solved, both in manufacturing and selling the cement. These difficulties, however, only made the proposition more inviting to him.

Edison followed his usual course of reading up all the literature on the subject that he could find, and seeking information from all quarters. After thorough study he came to the conclusion that with his improved methods of handling finely crushed material, and with some new inventions and processes he had in mind, he could go into the cement business and succeed in making a finer quality of product. As we shall see later, he "made good."

This study of the cement proposition took place during the first few months of his experimenting on a new storage battery. In the mean time Mr. Mallory had been busy arranging for the formation of a company with the necessary money to commence and carry on the business. One day he went to the laboratory and told Mr. Edison that everything was ready and that it was now time to engage engineers to lay out the works.

To this Edison replied that he intended to do that himself, and invited Mr. Mallory to go with him to one of the draughting-rooms up-stairs. Here Edison placed a sheet of paper on a draughting-table and immediately began to draw out a plan of the proposed works. He continued all day and away into the evening, when he finished; thus completing within twenty-four hours the full lay-out of the entire plant as it was subsequently installed. If the plant were to be rebuilt to-day no vital change would be necessary.

It will be granted that this was a remarkable engineering feat, for Edison was then a newcomer in the cement business. But in that one day's planning everything was considered and provided for, including crushing, mixing, weighing, grinding, drying screening, sizing, burning, packing, storing, and other processes.

From one end to the other the cement plant is about half a mile long, and through the various buildings there passes, automatically, each day a vast quantity of material under treatment. In practice this results in the production of more than two and a quarter million pounds of finished cement every twenty-four hours.

Not only was all this provided for in that one day's designing, but also smaller details, such, for instance, as the carrying of all steam, water and air pipes and electrical conductors in a large subway extending from one end of the plant to the other; also a system by which the ten thousand bearings in the plant are oiled automatically, requiring the services of only two men for the entire work.

Following this general outline plan of the whole plant by Edison himself there came the preparation of the detail plans by his engineers. As the manufacture of cement also involves the breaking and grinding of rocks, the scheme, of course, included using the giant rolls and other crushing, drying, and screening machinery invented by him for the iron-concentrating work, as mentioned in our last chapter.

No magnetic separator is necessary in cement-making, but there were other processes to provide for that did not occur in concentrating iron ore. One of them relates to burning the material, which is one of the most important processes in manufacturing cement.

Perhaps it may be well to state for the information of the reader that in cement-making, generally speaking, cement-rock and limestone in the rough are mixed together and ground to a fine powder. This powder is "burned" in a kiln and comes out in the form of balls, called "clinker." This again is crushed to a fine powder, which is the cement of commerce.

It will be seen, therefore, that the quantity of finished cement produced depends largely upon the capacity of the kilns. When Edison first thought of going into cement-making he expected to use the old style of kilns, which were about sixty feet long and six feet in diameter, and had a capacity of turning out about two hundred barrels of clinker every twenty-four hours. He is never satisfied, however, to take the experience of others as final, and thought he could improve on what had been done before.

He discussed the project with Mr. Mallory, who says: "After having gone over this matter several times, Mr. Edison said, 'I believe I can make a kiln which will give an output of one thousand barrels in twenty-four hours.' Although I had then been closely associated with him for ten years and was accustomed to see him accomplish great things, I could not help feeling the improbability of his being able to jump into an old-established industry—as a novice—and start by improving the 'heart' of the production so as to increase its capacity four hundred per cent. But Mr. Edison went to work immediately and very soon completed the design of a new type of kiln which was to be one hundred and fifty feet long and nine feet in diameter, made up in ten-foot sections of cast iron bolted together and arranged to be revolved on fifteen bearings. He had a wooden model made, and studied it very carefully through a series of experiments. These resulted so satisfactorily that this form was finally decided upon, and ultimately installed as part of the plant.

"Well, for a year or so the kiln problem was a nightmare to me. We could only obtain four hundred barrels at first, but gradually crept up through a series of heart-breaking trials until we got over eleven hundred barrels a day. Mr. Edison never lost his confidence throughout the trials, but on receiving a disappointing report would order us to try it again."

Although the older cement manufacturers predicted utter failure, they have since recognized the success of Edison's long kiln, and it is now being used quite generally in the trade.

Another invention of minor nature but worthy of note relates to the weighing of the proportions of cement-rock and limestone. In most cases the measurement is usually by barrow loads, but Edison determined that it must be done accurately to the pound, and devised a means of doing it automatically, for, as he remarked, "The man at the scales might get to thinking of the other fellow's best girl, so fifty or a hundred pounds of rock, more or less, wouldn't make much difference to him."

With Edison's device the scales are set at certain weights and the materials are fed from hoppers. The moment the scale-beam tips an electrical connection automatically stops the feed and no more can be put on the scale until the load is withdrawn.

Another and important new feature introduced by Edison was in raising the standard of fine grinding of cement ten points above the regular standard of seventy-five per cent, through a two-hundred-mesh screen. By reason of the great improvement he had made in grinding machinery he could grind cement so that eighty-five per cent, passed through a two-hundred-mesh screen. As cement is valuable in proportion to its fineness, it will be seen that he has thus made an advance of great importance to the trade.

We cannot enter into all the details of the numerous inventions and improvements that Edison has introduced into his cement plant during the last eight or nine years. It is sufficient to say that by his persistent and energetic labors during that period he has raised his plant from the position of a newcomer to the rank of the fifth largest producer of cement in this country.

A remarkable instance of the power of Edison's memory may be related here. Some years ago, when the cement plant was nearly finished and getting ready to start, he went up to look it over and see what needed to be done.

On the arrival of the train at ten-forty in the morning he went to the mill, and, starting at one end, went through the plant to the other end, examining every detail. He made no notes or memoranda, but the examination required all day.

In the afternoon, at five-thirty, he took a train for home, and on arriving there a few hours later got out some note-books and began to write from memory the things needing change or attention. He continued on this work all night and right along until the next afternoon, when he completed a list of nearly six hundred items. This memory "stunt" was the more remarkable because many of the items included all the figures of new dimensions he had decided upon for some of the machinery in the plant.

Each item was numbered consecutively, and the list copied and sent up to the superintendent, who was instructed to make the changes and report by number as they were done. These changes were made and their value was proven by later experience.

Edison's achievements have made a deep impression on the cement industry, but it is likely that it will become still deeper when his "Poured Cement House" is exploited.

A few years ago he conceived the idea of pouring a complete concrete house in a few hours. He made a long series of experiments for producing a free-flowing combination of the necessary materials, and at length found one that satisfied him that his idea was feasible, although experts said it could not be done.

His plan is to provide two sets of iron molds, one inside the other, with an open space between. These molds are made in small pieces and set up by being bolted together. When erected, the concrete mixture is poured in from the top in a continuous stream until the space between the molds is filled.

The pouring will be done in about six hours, after which the molds will be left in position about four days in order that the concrete may harden. When the molds are removed there will remain standing an entire house, complete from cellar to roof, with walls, floors, stairways, bath and laundry tubs, all in one solid piece. These houses, when built in quantity, can be produced at a very moderate cost.

Mr. Edison intends this house for the workingman, and in its design has insisted on its being ornamental as well as substantial. As he expressed it: "We will give the workingman and his family ornamentation in their house. They deserve it, and besides, it costs no more after the pattern is made to give decorative effects than it would to make everything plain."

XX

MOTION-PICTURES

Through his invention and introduction of the phonograph and of his apparatus for taking and exhibiting motion-pictures Edison has probably done more to interest and amuse the world than any other living man. These two forms of amusement have more audiences in a week than all the theaters in America in a year.

It is a curious fact that while instantaneous photography is necessary to produce motion pictures, the *suggestion* of producing them was made many years before the instantaneous photograph became possible.

One of the earliest efforts in this direction was made before Edison was born, and shown by a toy called the Zoetrope, or "Wheel of Life." A number of figures showing fractional parts of the motion of an object—such, for instance, as a boy skating—were boldly drawn in silhouette on a strip of paper. This paper was put inside an open cylinder having small openings around its circumference. The cylinder was mounted on a pivot, and, when revolved, the figures on the paper seemed to be in motion when viewed through the openings.

The success of this and similar toys, as well as of modern motion-pictures, depends upon a phenomenon known as the "persistence of vision." This means that if an object be presented to the vision for a moment and then withdrawn, the image of that object will remain impressed on the retina of the eye for a period of one-tenth to one-seventh of a second.

If, for instance, a bright light be moved rapidly up and down in front of the eye in a dark room it appears not as a single light, but as a line of fire, because there is not time for the eye to lose the image of the light between the rapid phases of its motion. For the same reason, if a number of pictures exactly alike were rapidly presented to the eye in succession it would seem as if a single picture were being viewed.

Thus, if a number of photographs, say at the rate of fifteen per second, be taken of a moving object, each successive photograph will show a fraction of the movements. Now if these photographs be thrown on a screen in the same order and at the same rate at which they were taken the movements of the object would apparently again take place, because the eye does not have time to lose the image of one fractional movement before the next follows.

One of the earliest suggestions of reproducing animate motion was made by a Frenchman named Ducos about 1864. He was followed by others, but they were all handicapped by the fact that dry-plates and sensitized film were entirely unknown, and the wet plates then used were entirely out of the question for the development of a practical commercial scheme.

The first serious attempt to secure photographs of objects in motion was made in 1878 by Edward Muybridge. At this time very rapid wet-plates were known. By arranging a line of cameras along a track and causing a horse in trotting past them to strike wires or strings attached to the shutters, the plates were exposed and a series of clear instantaneous photographs of the horse in motion was obtained.

Positive prints were made which were mounted in a modified form of Zoetrope and projected upon a screen. The horse in motion was thus reproduced, but, differing from the motion-pictures of to-day, always remained in the center of the screen in violent movement and making no progress.

Early in the 'eighties dry-plates were introduced, and other experimenters took up the work, but they were handicapped by the fact that plates were heavy and only a limited number could be used. This difficulty may be easily understood when it is realized that a modern motion-picture reel lasting fifteen minutes comprises about sixteen thousand separate and distinct photographs. The impossibility of manipulating this large number of glass plates to show one motion-picture play will be seen at once.

This was the condition of the art when Edison entered upon the work. He himself says, "In the year 1887 idea occurred to me that it was possible to devise an instrument which should do for the eye what the phonograph does for the ear, and that by a combination of the two all motion and sound could be recorded and reproduced simultaneously."

Two very serious difficulties lay in the way, however—first, a sensitive surface of such form and weight as could be successively brought into position and exposed at a very high rate; and, secondly, the making of a camera capable of so taking the pictures. Edison proved equal to the occasion, and, after an immense amount of work and experiment, continuing over a long period of time, succeeded in producing apparatus that made modern motion-pictures possible.

In his earliest experiments a cylinder about the size of a phonograph record was used. It was coated with a highly sensitized surface, and microscopic photographs, arranged spirally, were taken upon it. Positive prints were made in the same way, and viewed through a magnifying-glass. Various forms of this apparatus were made, but all were open to serious objections, the chief trouble being with the photographic emulsion.

During this experimental period the kodak film was being developed by the Eastman Kodak Company, under the direction of Mr. George Eastman. Edison recognized that in this product there lay the solution of that part of the problem. At first the film was not just what he required, but the Eastman Company after a time developed and produced the highly sensitized surface that Edison sought.

It then remained to devise a camera by means of which from twenty to forty pictures per second could be taken. Every user of a film camera can appreciate the difficulty of the problem. A long roll of film must pass steadily behind the lens. At every inch it must be stopped, the shutter opened for the exposure, and then closed again. The film must be advanced say an inch, and these operations repeated twenty to forty times a second throughout, perhaps, a thousand feet of film.

Who but an Edison would assume that such a device could be made, and with such exactness that each picture should coincide with the others? After much experiment, however, he finally accomplished it, and in the summer of 1889 the first modern motion-picture camera was made. From that day to this the Edison camera has been the accepted standard for securing pictures of objects in motion.

The earliest form of exhibiting apparatus was known as the kinetoscope. It was a machine in which a positive print from the negative roll of film obtained in the camera was exhibited directly to the eyes through a peep-hole. About

1895 the pictures were first shown through a modified form of magic lantern, and have so continued to this day. The industry has grown very rapidly, and for a long time the principal American manufacturers of motion-pictures paid a royalty to Edison under his basic patents.

The pictures made in the earliest days of the art were simple and amusing, such as Fred Ott's sneeze, Carmencita dancing, Italians and their performing bears, fencing, trapeze stunts, horsemanship, blacksmithing, and so on. No attempt was made to portray a story or play. The "boys" at the laboratory laugh when they tell of a local bruiser who agreed to box a few rounds with "Jim" Corbett in front of the camera. When this local "sparring partner" came to face Corbett he was so paralyzed with terror he could hardly move.

These early pictures were made in the yard of Edison's laboratory at Orange, in a studio called the "Black Maria." It was made of wood, painted black inside and out, and could be swung around to face the sunlight, which was admitted by a movable part of the roof.

This is all very different in these modern days. The studios in which interior motion-pictures are made are expensive and pretentious affairs. An immense building of glass, with all the properties and stage settings of a regular theater, are required. Of course many of the plays are produced out of doors, in portions of the country suited to the story.

All the companies producing motion-pictures employ regular stock companies of actors and actresses, selected especially for their skill in pantomime, although, as may be suspected, in the actual taking of the pictures they are required to carry on an animated dialogue as if performing on the real stage. This adds to the smoothness and perfection of the performance.

Motion-picture plays are produced under the direction of skilled stage-managers who must be specially trained for this particular business. Their work is far from being easy, for an act in a picture-play must be exact and free from mistakes, and must take place in a very short time. For instance, an act in such a play may take less than five minutes to perform, but it must be carefully rehearsed for several weeks beforehand.

There is plenty of scope for patience and ingenuity in taking motion-picture plays. If trained children or animals are required they must be found or trained; and all the resources of trick and stop photography are called upon from time to time as the occasion requires.

Edison has always held to his idea of a combination of the phonograph and motion-picture. Some time ago he said, "I believe that in coming years, by my own work and that of Dickson, Muybridge, Marey, and others who will doubtless enter the field, grand opera can be given at the Metropolitan Opera House in New York without any material change from the original, and with artists and musicians long since dead."

This prediction has been partly fulfilled, for Edison's successful talking motion-pictures marked the beginning of the "talkies" which are flourishing to-day.

XXI

EDISON INVENTS A NEW STORAGE BATTERY

Many an invention has been made as the result of some happy thought or inspiration, but most inventions are made by men working along certain lines, who set out to accomplish a desired result. It is rarely, however, that man starts out deliberately, as Edison did, to invent an entirely new type of such an intricate device as a storage battery, with only a vague starting point.

Previous to Edison's work the only type of storage battery known was the one in which lead plates and sulphuric acid were employed. He had always realized the value of a storage battery as such, but never believed that the lead-acid type could fulfil all expectations because of its weight and incurable defects.

About the time that he closed the magnetic iron ore concentrating plant (in the beginning of the present century) Edison remarked to Mr. R. H. Beach, then of the General Electric Company: "Beach, I don't think nature would be so unkind as to withhold the secret of a *good* storage battery if a real earnest hunt for it is made. I'm going to hunt." And before starting he determined to avoid lead and sulphuric acid.

Edison is frequently asked what he considers to be the secret of achievement. He always replies, "Hard work, based on hard thinking." He has consistently lived up to this prescription to the utmost.

Of all his inventions it is doubtful whether any one of them has called forth more original thought, work, perseverance, ingenuity, and monumental patience than the one we are now dealing with. One of his associates who has been through the many years of the storage-battery drudgery with him said: "If Edison's experiments, investigations, and work on this storage battery were all that he had ever done, I should say that he was not only a notable inventor, but also a great man. It is almost impossible to appreciate the enormous difficulties that have been overcome."

From a beginning which was made practically in the dark, it was not until he had completed more than ten thousand experiments that he obtained any positive results whatever. Month after month of constant work by day and night had not broken down Edison's faith in success, and the failure of an experiment simply meant that he had found something else that would *not* do, thus bringing him nearer the possible goal.

After this immense amount of preliminary work he had obtained promising results in a series of reactions between nickel and iron, and was then all afire to push ahead. He therefore established a chemical plant at Silver Lake, New Jersey, and, gathering around him a corps of mechanics, chemists, machinists, and experimenters, settled down to one of his characteristic struggles for supremacy. To some extent it was a revival of the old Menlo Park days and nights.

The group that took part in these early years of Edison's arduous labors included his old-time assistant, Fred Ott, together with his chemist, J. W. Aylsworth, as well as E. J. Ross, Jr.; W. E. Holland, and Ralph Arbogast, and a little later W. G. Bee, all of whom grew up with the battery and devoted their energies to its commercial development.

One of these workers, relating the strenuous experiences of these few years, says: "It was hard work and long hours, but still there were some things that made life pleasant. One of them was the supper-hour we enjoyed when we worked nights. Mr. Edison would have supper sent in about midnight, and we all sat down together, including himself. Work was forgotten for the time, and all hands were ready for fun. I have very pleasant recollections of Mr. Edison at these times. He would always relax and help to make a good time, and on some occasions I have seen him fairly overflow with animal spirits, just like a boy let out of school. He was very fond of telling and hearing stories, and always appreciated a joke. After the supper-hour was over, however, he again became the serious, energetic inventor, deeply immersed in the work in hand."

Another interesting and amusing reminiscence of this period of activity has been told by another of the family of experimenters: "Sometimes when Mr. Edison had been working long hours he would want to have a short sleep. It was one of the funniest things I ever witnessed to see him crawl into an ordinary roll-top desk and curl up and take a nap. If there was a sight that was still more funny, it was to see him turn over on his other side, all the time remaining in the desk. He would use several volumes of *Watts' Dictionary of Chemistry* for a pillow, and we fellows used to say that he absorbed the contents during his sleep, judging from the flow of new ideas he had on waking."

Such incidents as these serve merely to illustrate the lighter moments that relieved the severe and arduous labors of the strenuous five years of the early storage-battery work of Edison and his associates. Difficulties there were a-plenty, but these are what Edison usually thrives on. As another coworker of this period says:

"Edison seemed pleased when he used to run up against a serious difficulty. It would seem to stiffen his backbone and make him more prolific of new ideas. For a time I thought I was foolish to imagine such a thing, but I could never get away from the impression that he really appeared happy when he ran up against a serious snag."

It would be out of the question in a book of this kind to follow Edison's trail in detail through the innumerable twists and turns of his experimentation on the storage battery, for they would fill a big volume. The reader may imagine how extensive they were from the reply of one of his laboratory assistants, who, when asked how many experiments were made on the storage battery since the year 1900, replied: "Goodness only knows! We used to number our experiments consecutively from one to ten thousand, and when we got up to ten thousand we turned back to one and ran up to ten thousand again, and so on. We ran through several series—I don't know how many, and have lost track of them now, but it was not far from fifty thousand."

The mechanical problems in devising this battery were numerous and intricate, but the greatest difficulty that Edison had to overcome was the proper preparation of nickel hydrate for the positive and iron oxide for the negative plate. He found that comparatively little was known by manufacturing chemists about these compounds. Hence it became necessary for him to establish his own chemical works and put them in charge of men specially trained by himself.

After an intense struggle with these problems, lasting over several years, the storage battery was at length completed and put on the market. The public was ready for it and there was a rapid sale.

Continuous tests of the battery were carried on at the laboratory, as well as practical and heavy tests in automobiles, which were kept running constantly over all kinds of roads under Edison's directions. After these tests had been going on for some time the results showed that occasionally a cell here and there would fall short in capacity.

This did not suit Edison. He was determined to make his storage battery a complete success, and after careful thought decided to shut down until he had overcome the trouble. The customers were satisfied and wanted to buy more batteries, but he was not satisfied and would sell no more until he had made the battery perfect.

He therefore shut down the factory and went to experimenting once more. The old strenuous struggle set in and continued nearly three years before he was satisfied beyond doubt that the battery was right. In the early summer of 1909 Edison once more started to manufacture and sell the batteries, and has since that time continued to supply them as quickly as they are made. At the present writing the factory is running day and night in attempting to keep up with orders.

One of the principal troubles of the earlier cells was a lack of conductivity between the nickel hydrate and the metal tube in which it was contained. Edison had used graphite to obtain this conductivity, but this material proved to be uncertain in some cases. After a long course of study and experiment he solved this problem in a satisfactory manner by using flakes of pure nickel, which he obtained by a most fascinating and ingenious process.

A metallic cylinder is electroplated with alternate layers of copper and nickel, one hundred of each. The combined sheet, which is only as thick as a visiting-card, is stripped off the cylinder and cut into tiny squares of about one-sixteenth of an inch each. These squares are put into a bath which dissolves out the copper. This releases the layers of nickel, so that each of these squares becomes one hundred tiny sheets, or flakes, of pure metallic nickel, so thin and light that when they are dried they will float in the air. These flakes are automatically pressed into the positive tubes with the nickel hydrate in an ingenious machine which had to be specially invented for the purpose.

Not only was this machine specially invented, but it was necessary to invent and design practically all the other machinery that it was necessary to use in manufacturing the battery. Thus, we see that in this, as in many other of Edison's inventions, it is not only the thing itself that has been invented, but also the special machinery and tools to make it.

The principal use that Edison has had in mind for his storage battery is the transportation of freight and passengers by truck, automobile, and street-car. Although at the time of writing this book the improved battery has been on the market a little over two years, great strides have been made in carrying his ideas into effect.

The number of trucks and automobiles using Edison's storage battery already run into the thousands, with more orders than can be immediately filled.

XXII

EDISON'S MISCELLANEOUS INVENTIONS

Thus far the history of Edison's career has fallen naturally into a series of chapters each aiming to describe a group of inventions in the development of some art. This plan has been helpful to the writer and probably useful to the reader.

It happens, however, that the process has left a vast mass of discovery and invention untouched, and it is now proposed to make brief mention of a few of the hundreds of things that have occupied Edison's attention from time to time.

Beginning with telegraphy, we find that Edison did some work on wireless transmission. He says: "I perfected a system of train telegraphy between stations and trains in motion, whereby messages could be sent from the moving train to the central office; and this was the forerunner of wireless telegraphy. This system was used for a number of years on the Lehigh Valley Railroad on their construction trains. The electric wave passed from a piece of metal on top of the car across the air to the telegraph wires, and then proceeded to the despatcher's office. In my first experiments with this system I tried it on the Staten Island Railroad and employed an operator named King to do the experimenting. He reported results every day, and received instructions by mail; but for some reason he could send messages all right when the train went in one direction, but could not make it go in the contrary direction. I made suggestions of every kind to get around this phenomenon. Finally I telegraphed King to find out if he had any suggestions himself, and I received a reply that the only way he could propose to get around the difficulty was to put the island on a pivot so it could be turned around. I found the trouble finally, and the practical introduction on the Lehigh Valley road was the result. The system was sold to a very wealthy man, and he would never sell any rights or answer letters. He became a spiritualist subsequently, which probably explains it."

The earlier experiments with wireless telegraphy were made at Menlo Park during the first days of the electric light, and it was not until 1886 that Edison had time to spare to put the system into actual use. At that time Ezra T. Gilliland and Lucius J. Phelps, who had experimented on the same lines, became associated with him in the work.

Although the space between the train and the pole line was not more than fifty feet, Edison had succeeded at Menlo Park in transmitting messages through the air at a distance of five hundred and eighty feet. Speaking of this and of his other experiments with induction telegraphy by means of kites, he said, recently: "We only transmitted about two and one-half miles through the kites. What has always puzzled me since is that I did not think of using the results of my experiments on 'etheric force' that I made in 1875. I have never been able to understand how I came to overlook them. If I had made use of my own work I should have had long-distance wireless telegraphy."

These experiments of 1875, as recorded in Edison's famous note-books, show that in that year he detected and studied some then unknown and curious phenomena which made him think he was on the trail of a new force. His representative, Mr. Batchelor, showed these experiments with Edison's apparatus, including the "dark box," at the

Paris Exposition in 1881. Without knowing it, for he was far in advance of the time, Edison had really entered upon the path of long-distance wireless telegraphy, as was proven later when the magnificent work of Hertz was published.

When Roentgen made the discovery of the X-ray in 1895 Edison took up experimentation with it on a large scale. He made the first fluoroscope, using tungstate of calcium for the screen. In order to find other fluorescent substances he set four men to work and thus collected upward of eight thousand different crystals of various chemical combinations, of which about eighteen hundred would fluoresce to the X-ray. He also invented a new lamp for giving light by means of these fluorescent crystals fused to the inside of the glass. Some of these lamps were made and used for a time, but he gave up the idea when the dangerous nature of the X-ray became known.

It would be possible to go on and describe in brief detail many more of the hundreds of Edison's miscellaneous inventions, but the limits of our space will not permit more than the mere mention of a *few*, simply to illustrate the wide range of his ideas and work. For instance:

- A dry process of separating placer gold; the rapid disposal of heavy snows in cities.

- Experiments on flying machines with an engine operated by explosions of guncotton.

- The joint invention, with M. W. Scott Sims, of a dirigible submarine torpedo operated by electricity.

- Pyromagnetic generators for generating electricity directly from the combustion of coal.

- Pyromagnetic motors operated by alternate heating and cooling.

- A magnetic bridge for testing the magnetic qualities of iron.

- A "dead-beat" galvanometer without coils or magnetic needle.

- The odoroscope, for measuring odors; preserving fruit *in vacuo*; making plate glass; drawing wire.

- Metallurgical processes for treatment of nickel, gold, and copper ores.

From first to last Edison has filed in the United States Patent Office more than fourteen hundred applications for patents. Besides, he filed some one hundred and twenty caveats, embracing not less than fifteen hundred additional inventions. The caveat has now been abolished in patent-office practice, but such a document could formerly be filed by an inventor to obtain a partial protection for a year while completing his invention. As an example of Edison's fertility and the endless variety of subjects engaging his attention the following list of matters covered by *one* of his caveats is given. All his caveats are not quite so full of "plums," but this is certainly a wonder:

- Forty-one distinct inventions relating to the phonograph, covering various forms of recorders, arrangement of parts, making of records, shaving tool, adjustments, etc.

- Eight forms of electric lamps using infusible earthy oxides and brought to high incandescence *in vacuo* by high potential current of several thousand volts; same character as impingement of X-rays on object in bulb.

- A loud-speaking telephone with quartz cylinder and beam of ultra-violet light.

- Four forms of arc-light with special carbons.

- A thermostatic motor.

- A device for sealing together the inside part and bulb of an incandescent lamp mechanically.

- Regulators for dynamos and motors.

- Three devices for utilizing vibrations beyond the ultra-violet.

- A great variety of methods for coating incandescent lamp filaments with silicon, titanium, chromium, osmium, boron, etc.

- Several methods of making porous filaments.

- Several methods of making squirted filaments of a variety of materials, of which about thirty are specified.

- Seventeen different methods and devices for separating magnetic ores.

- A continuously operative primary battery.

- A musical instrument operating one of Helmholtz's artificial larynxes.

- A siren worked by explosion of small quantities of oxygen and hydrogen mixed.

- Three other sirens made to give vocal sounds or articulate speech.

- A device for projecting sound-waves to a distance without spreading, and in a straight line, on the principle of smoke-rings.

- A device for continuously indicating on a galvanometer the depths of the ocean.

- A method of preventing in a great measure friction of water against the hull of a ship and incidentally preventing fouling by barnacles.

- A telephone receiver whereby the vibrations of the diaphragm are considerably amplified.

- Two methods of "space" telegraphy at sea.

- An improved and extended string telephone.

- Devices and method of talking through water for a considerable distance.

- An audiphone for deaf people.

- Sound-bridge for measuring resistance of tubes and other materials for conveying sound.

- A method of testing a magnet to ascertain the existence of flaws in the iron or steel composing the same.

- Method of distilling liquids by incandescent conductor immersed in the liquid.

- Method of obtaining electricity direct from coal.

- An engine operated by steam produced by the hydration and dehydration of metallic salts.

- Device and method of telegraphing photographically.

- Carbon crucible kept brilliantly incandescent by current *in vacuo* for obtaining reaction with refractory metals.

- Device for examining combinations of odors and their changes by rotation at different speeds.

It must be borne in mind that the above and hundreds of others are not merely *ideas* put in writing, but represent actual inventions upon which Edison worked and experimented. In many cases the experiments ran into the thousands, requiring months for their performance.

To describe Edison's mere ideas and suggestions for future work would of itself fill a volume. These are written in his own handwriting in a number of large record-books which he has shown to the writer. Judging from a hasty inspection, there is enough material in these books to occupy the lifetime of several persons.

The immense range of Edison's mind and activities cannot well be described in cold print, but can only be adequately comprehended by those who have been closely associated with him for a length of time, and who have had opportunity of studying his voluminous records.

XXIII

EDISON'S METHOD IN INVENTING

If one were allowed only two words with which to describe Edison it is doubtful whether a close examination of the entire dictionary would disclose any others more suitable than "experimenter-inventor." These would express the overruling characteristics of his eventful career.

His life as child, boy, and man has revealed the born investigator with original reasoning powers, unlimited imagination, and daring method. It is not surprising, therefore, that a man of this kind should exhibit a ceaseless, absorbing desire for knowledge, willing to spend his last cent in experimentation to satisfy the cravings of an inquiring mind.

There is nothing of the slap-dash style in Edison's experiments. While he "tries everything," it is not merely the mixing of a little of this, some of that, and a few drops of the other, in the *hope* that *something* will come of it. On the contrary, his instructions are always clear-cut and direct, and must be followed out systematically, exactly, and minutely, no matter where they lead nor how long the experiment may take.

Unthinking persons have had a notion that some of Edison's successes have been due to mere dumb fool luck—to fortunate "happenings." Nothing could be farther from the truth, for, on the contrary, it is owing almost entirely to his comprehensive knowledge, the breadth of his conception, the daring originality of his methods, and minuteness and extent of experiment, combined with patient, unceasing perseverance, that new arts have been created and additions made to others already in existence.

One of the first things Edison does in beginning a new line of investigation is to master the literature of the subject. He wants to know what has been done before. Not that he considers this as final, for he often obtains vastly different results by repeating in his own way the experiments of others.

"Edison can travel along a well-used road and still find virgin soil," remarked one of his experimenters recently, who had been trying to make a certain compound, but with poor success. Edison tried it in the same way, but made a change in one of the operations and succeeded.

Another of the experimental staff says: "Edison is never hindered by theory, but resorts to actual experiment for proof. For instance, when he conceived the idea of pouring a complete concrete house it was universally held that it would be impossible because the pieces of stone in the mixture would not rise to the level of the pouring-point, but would gravitate to a lower plane in the soft cement. This, however, did not hinder him from making a series of experiments which resulted in an invention that proved conclusively the contrary."

Having conceived some new idea and read everything obtainable relating to the subject in general, Edison's fertility of resource and originality come into play. He will write in one of the laboratory note-books a memorandum of the experiments to be tried, and, if necessary, will illustrate by sketches.

This book is then given to one of the large staff of experimenters. Here strenuousness and a prompt carrying on of the work are required. The results of each experiment must be recorded in the notebook, and daily or more frequent reports are expected. Edison does not forget what is going on, but in his daily tours through the laboratory keeps in touch with the work of all the experimenters. His memory is so keen and retentive that he is as fully aware of the progress and details of each of the numerous experiments constantly going on as if he had made them all himself.

The use of laboratory note-books was begun early in the Menlo Park days and has continued ever since. They are plain blank-books, each about eight and a half by six inches, containing about two hundred pages. At the present time there are more than one thousand of these books in the series. On their pages are noted Edison's ideas, sketches, and memoranda, together with records of countless thousands of experiments made by him or under his direction during more than thirty years.

These two hundred thousand or more pages cover investigations into every department of science, showing the operations of a master mind seeking to surprise Nature into a betrayal of her secrets by asking her the same question in a hundred different ways. The breadth of thought, thoroughness of method, infinite detail, and minuteness of investigation proceeding from the workings of one mind would surpass belief were they not shown by this wonderful collection of note-books.

A remark made by one of the staff, who has been experimenting at the laboratory for over twenty years, is suggestive. He said: "Edison can think of more ways of doing a thing than any man I ever saw or heard of. He tries everything and never lets up, even though failure is apparently staring him in the face. He only stops when he simply can't go any farther on that particular line. When he decides on any mode of procedure he gives his notes to the experimenter and lets him alone, only stopping in from time to time to look at the operations and receive reports of progress."

The idea of attributing great successes to "genius" has always been repudiated by Edison, as evidenced by his historic remark that "genius is one per cent, inspiration and ninety-nine per cent, perspiration." Again, in a conversation many years ago between Edison, Batchelor, and E. H. Johnson, the latter made allusion to Edison's genius, when Edison replied:

"Stuff! I tell you genius is hard work, stick-to-it-iveness, and common sense."

"Yes," said Johnson, "I admit there is all that to it, but there's still more. Batch and I have those qualifications, but, although we knew quite a lot about telephones, and worked hard, we couldn't invent a brand-new non-infringing telephone receiver as you did when Gouraud cabled for one. Then, how about the subdivision of the electric light?"

"Electric current," corrected Edison.

"True," continued Johnson; "you were the one to make that very distinction. The scientific world had been working hard on subdivision for years, using what appeared to be common sense. Results, worse than nil. Then you come along, and about the first thing you do, after looking the ground over, is to start off in the opposite direction, which subsequently proves to be the only possible way to reach the goal. It seems to me that this is pretty close to the dictionary definition of genius."

It is said that Edison replied rather incoherently and changed the topic of conversation.

This innate modesty, however, does not prevent Edison from recognizing and classifying his own methods of investigation. In a conversation with two old associates a number of years ago he remarked: "It has been said of me that my methods are empirical. That is true only so far as chemistry is concerned. Did you ever realize that practically all industrial chemistry is colloidal in its nature? Hard rubber, celluloid, glass, soap, paper, and lots of others, all have to deal with amorphous substances, as to which comparatively little has been really settled. My methods are similar to those followed by Luther Burbank. He plants an acre, and when this is in bloom he inspects it. He has a sharp eye, and can pick out of thousands a single plant that has promise of what he wants. From this he gets the seed, and uses his skill and knowledge in producing from it a number of new plants which, on development, furnish the means of propagating an improved variety in large quantity. So, when I am after a chemical result that I have in mind I may make hundreds or thousands of experiments out of which there may be one that promises results in the right direction. This I follow up to its legitimate conclusion, discarding the others, and usually get what I am after. There is no doubt about this being empirical; but when it comes to problems of a mechanical nature, I want to tell you that all I've ever tackled and solved have been done by hard, logical thinking." The intense earnestness and emphasis with which this was said were very impressive to the auditors.

If, in following out his ideas, an experiment does not show the results that Edison wants, it is not regarded as a failure, but as something learned. This attitude is illustrated by his reply to Mr. Mallory, who expressed regret that the first nine thousand and odd experiments on the storage battery had been without results. Edison replied, with a smile:

"Results! Why, man, I have gotten a lot of results! I have found several thousand things that won't work."

Edison's patient, plodding methods do not always appear on the note-books. For instance, a suggestion in one of them refers to a stringy, putty-like mass being made of a mixture of lampblack and tar. Some years afterward one of the laboratory assistants was told to make some and roll it into filaments. After a time he brought the mass to Edison and said:

"There's something wrong about this, for it crumbles even after manipulating it with my fingers."

"How long did you knead it?" asked Edison.

"Oh, more than an hour," was the reply.

"Well, keep on for a few hours more and it will come out all right," was the rejoinder. And this proved to be correct.

With the experimenter or employee who exercises thought Edison has unbounded patience, but to the careless, stupid, or lazy person he is a terror for the short time they remain around him. Once, when asked why he had parted with a certain man, he said: "Oh, he was so slow that it would take him half an hour to get out of the field of a microscope."

Edison's practical way of testing a man's fitness for special work is no joke, according to Mr. J. H. Vail, formerly one of the Menlo Park staff. "I wanted a job," he said, "and was ambitious to take charge of the dynamo-room. Mr. Edison led me to a heap of junk in a corner and said: 'Put that together and let me know when it is running.' I didn't know what it was, but received a liberal education in finding out. It proved to be a dynamo, which I finally succeeded in assembling and running. I got the job."

A somewhat similar experience is related by Mr. John F. Ott, who, in 1869, applied for work. This is the conversation that took place, led by Edison's question:

"What do you want?"

"Work."

"Can you make this machine work?" (Exhibiting it and explaining its details).

"Yes."

"Are you sure?"

"Well, you needn't pay me if I don't."

And thus Mr. Ott went to work and accomplished the results desired. Two weeks afterward Edison put him in charge of the shop. From that day to this, Mr. Ott has remained a member of Mr. Edison's staff.

Examples without number could be given of Edison's inexhaustible fund of ideas, but one must suffice by way of example. In the progress of the ore-concentrating work one of the engineers submitted three sketches of a machine for some special work. They were not satisfactory. He remarked that it was too bad there was no other way to do the work. Edison said, "Do you mean to say that these drawings represent the only way to do this work?" The reply was, "I certainly do." Edison said nothing, but two days afterward brought in his own sketches showing *forty-eight* other ways of accomplishing the result, and laid them on the engineer's desk without a word. One of these ideas, with slight changes, was afterward adopted.

This chapter could be continued to great length, but must now be closed in the hope that in the foregoing pages the reader may have caught an adequate glance of Mr. Edison at work.

XXIV

EDISON'S LABORATORY AT ORANGE

If Longfellow's youth "Who through an Alpine village passed" had been Edison, the word upon his banner would probably not have been "Excelsior" but "Experiment." This seems to be the watchword of his life, and is well illustrated by a remark made to Mr. Mason, the superintendent of the cement works: "You must experiment all the time; if you don't the other fellow will, and then he will get ahead of you."

For some years after closing the little laboratory in his mother's cellar Edison made a laboratory of any nook or corner and experimented as long as he had a dollar in his pocket. The first place he began to do larger things was in Newark, where he established his first shops.

While life there was very strenuous, he tells of some amusing experiences: "Some of my assistants in those days were very green in the business. One day I got a new man and told him to conduct a certain experiment. He got a quart of ether and started to boil it over a naked flame. Of course it caught fire. The flame was about four feet in diameter and eleven feet high. The fire department came and put a stream through the window. That let all the fumes and chemicals out and overcame the firemen.

"Another time we experimented with a tubful of soapy water and put hydrogen into it to make large bubbles. One of the boys, who was washing bottles in the place, had read in some book that hydrogen was explosive, so he proceeded to blow the tub up. There was about four inches of soap in the bottom of the tub, which was fourteen inches high, and he filled it with soap-bubbles up to the rim. Then he took a bamboo fish-pole, put a piece of lighted paper at the end and touched it off. It blew every window out of the place."

We have seen that Edison moved to Menlo Park, where he had a very complete laboratory, in which he brought out a large number of important inventions. After a time, however, this establishment was outgrown and lost many of its possibilities, and he began to plan a still greater one which should be the most complete of its kind in the world.

The Orange laboratory, as was originally planned, consisted of a main building two hundred and fifty feet long and three stories in height, together with four other structures, each one hundred by twenty-five feet and only one story in height. All these were substantially built of brick. The main building was divided into five chief divisions—the library, office, machine-shops, experimental and chemical rooms, and stock-rooms. The small buildings were to be used for various purposes.

A high picket fence, with a gate, surrounded these buildings. A keeper was stationed at the gate with instructions to admit no strangers without a pass. On one occasion a new gateman was placed in charge, and, not knowing Edison, refused to admit him until he could get some one to come out and identify him.

The library is a spacious room about forty by thirty-five feet. Around the sides of the room run two tiers of gallery. The main floor and the galleries are divided into alcoves, in which, on the main floor, are many thousands of books.

In the galleries are still more books and periodicals of all kinds, also cabinets and shelves containing mineralogical and geological specimens and thousands of samples of ores and minerals from all parts of the world. In a corner of one of the galleries may be seen a large number of magazines relating to electricity, chemistry, engineering, mechanics, building, cement, building materials, drugs, water and gas power, automobiles, railroads, aeronautics, philosophy, hygiene, physics, telegraphy, mining, metallurgy, metals, music, and other subjects; also theatrical weeklies, as well as the proceedings and transactions of various learned and technical societies. All of these form part of Mr. Edison's current reading. At one end of the main floor of the library, which is handsomely and comfortably furnished, is Mr. Edison's desk, at which he may usually be seen for a while in the early morning hours or at noon looking over his mail.

The centre of the library is left open for the reception of visitors, and one corner is partitioned off to provide a private office for Mr. Edison's son, Charles, who is the President and active manager of the various Edison industries. Directly opposite to the entrance-door is a beautiful marble statue representing the supremacy of electric light over gas. This statue was purchased by Mr. Edison at the Paris Exposition in 1889.

A glance at the book-shelves affords a revelation of the subjects in which Edison is interested, for the titles of the volumes include astronomy, botany, chemistry, dynamics, electricity, engineering, forestry, geology, geography, mechanics, mining, medicine, metallurgy, magnetism, philosophy, psychology, physics, steam, steam-engines, telegraphy, telephony, and many others. These are not all of Edison's books by any means, for he has another big library in his house on the hill.

Turning to pass out of the library, one's attention is arrested by a cot standing in one of the alcoves near the door. Sometimes during long working hours Mr. Edison will throw himself down for a nap. He has the ability to go to sleep instantly, and, being deaf, noises do not disturb his slumber. The instant he awakes he is in full possession of his faculties and goes "back to the job" without a moment's hesitation.

Immediately outside the library is the famous stock-room, about which much has been written. Edison planned to have in this stock-room some quantity, great or small, of every known substance not easily perishable, together with the most complete assortment of chemicals and drugs that experience and knowledge could suggest. His theory was, and is, that he does not know in advance what he may want at any moment, and he planned to have anything that could be thought of ready at hand.

Thus, the stock-room is not only a museum, but a sample-room of nature, as well as a supply department. At first glance the collection is bewildering, but when classified is more easily comprehended.

The classification is natural, as, for instance, objects pertaining to various animals, birds, and fishes, such as skins, hides, hair, fur, feathers, wool, quills, down, bristles, teeth, bones, hoofs, horns, tusks, shells; natural products such as woods, barks, roots, leaves, nuts, seeds, gums, grains, flowers, meals, bran; also minerals in great assortment; mineral and vegetable oils, clay, mica, ozokerite, etc. In the line of textiles, cotton and silk threads in great variety, with woven goods of all kinds, from cheese-cloth to silk plush. As for paper, there is everything in white and color, from thinnest tissue up to the heaviest asbestos, even a few newspapers being always on hand. Twines of all sizes, inks, wax, cork, tar, rosin, pitch, asphalt, plumbago, glass in sheets and tubes, and a host of miscellaneous articles are revealed on looking around the shelves, as well as an interminable collection of chemicals including acids, alkalies, salts, reagents, every conceivable essential oil, and all the thinkable extracts. It may be remarked that this collection includes the eighteen hundred or more fluorescent salts made by Edison during his experiments for the best material for a fluoroscope in the early X-ray period. All known metals in form of sheet, rod, and tube, and of great variety in thickness, are here found also, together with a most complete assortment of tools and accessories for machine-shop and laboratory work.

The list above given is not by any means complete. In detail it would stretch out to a rather large catalogue. It is not by any means an idle collection, for a stock clerk is kept busy all the day answering demands upon him.

Beyond the stock-room is a good-sized machine-shop, well equipped, in which the heavier class of models and mechanical devices are made. Attached to these are the engine-room and boiler-room. Above, on the second floor, is another machine-shop, in which is carried on work of greater precision and fineness in the construction of tools and experimental models.

There are many experimental rooms on the second and third floors of the laboratory building. In these the various experimenters are at work carrying out the ideas of Mr. Edison on the great variety of subjects to which he is constantly devoting his attention. One cannot go far in the upper floors without being aware that efforts are being made to improve the phonograph, for the sounds of vocal and instrumental music can be heard from all sides.

On the third floor the visitor may see a number of glass-fronted cabinets containing a multitude of experimental incandescent lamps, and an immense variety of models of phonographs, motors, telegraph and telephone apparatus, and a host of other inventions, upon which Mr. Edison's energies have at one time or other been bent. Here are also many boxes of historical instruments and models. In fact, this hallway, with its variety of contents, may well be considered a scientific attic.

In the early days of the Orange laboratory some of the upper rooms contained cots for the benefit of the night-workers. In spite of the strenuous nights and days the spirit of fun was frequently in evidence. One instance will serve as an illustration.

One morning about two-thirty the late Charles Batchelor said he was tired and would go to bed. Leaving Edison and the others busily working, he went out and returned quietly in slippered feet, with his night-gown on, the handle of a feather-duster down his back with the feathers waving over his head, and his face marked. With unearthly howls and shrieks, *a l'Indien*, he pranced about the room, incidentally giving Edison a scare that made him jump up from his work. He saw the joke quickly, however, and joined in the general merriment caused by this prank.

A description of the laboratory building would be incomplete without mention of room Number 12. This is one of Edison's favorite rooms, where he may frequently be found seated at a plain table in the center of the room deeply intent on one of his numerous problems. It is a plain little room, but seems to exercise a nameless fascination for him.

Passing out of the building, we come to the four smaller buildings, which are known as Numbers One, Two, Three, and Four. The building Number One is called the galvanometer room. Edison originally planned that this should be used for the most delicate and minute electrical measurements. He went to great expense in fitting it up and in providing a large number of costly instruments, but the coming of the trolley near by a few years afterward rendered the room utterly useless for this purpose. It is now used as an experimental room, chiefly for motion-picture experiments.

Building Number Two is quite an important one. As the visitor arrives at the door he is quite conscious that it is a chemical-room. Here a corps of chemists is constantly kept busy in carrying out the various experiments Mr. Edison has given them to perform. This room is also one of his special haunts. He may be seen here very frequently experimenting in person, or seated at a plain little table figuring out some new combination that he has in mind.

A chemical store-room and a pattern-maker's shop occupy building Number Three, while Number Four, which was formerly used for ore concentrating experiments, is now used as a general stock-room.

We have only attempted to afford the reader a passing glance of this interesting laboratory, which for many years has been the headquarters of Edison and the central source of inspiration for the great industries he has established at Orange. Around it are grouped a number of immense concrete buildings in which the manufacture of phonographs, motion-pictures, and storage batteries is carried on, giving employment to as many as four thousand people during busy times.

Needless to say, the laboratory has many visitors. Celebrities of all kinds and distinguished foreigners are numerous, coming from all parts of the world to see the great inventor and the scene of his activities.

The Boy's Life of Edison

XXV

EDISON HIMSELF

Let us turn from what Edison has done to what Edison is. It is worth while to know "the man behind the guns." Who and what is the personal Edison?

Certainly there must be tremendous force in a personality which has been one of the most potent factors in bringing into existence new industries now capitalized at tens of billions of dollars, earning annually sums running into billions, and giving employment to an army of more than two million people.

It must not be thought that there is any intention to give entire credit to Edison for the present magnificent proportions of these industries. The labors of many other inventors and the confidence of capitalists and investors have added greatly to their growth. But Edison is the father of some of these arts and industries, and as to some of the others it was the magic of his touch that helped make them practicable.

How then does Edison differ from most other men? Is it that he combines with a vigorous body a mind capable of clear and logical thinking, and an imagination of unusual activity? No, for there are others of equal bodily and mental vigor who have not accomplished a tithe of his achievements.

We must answer then, first, that his whole life is concentrated upon his work. When he conceives a broad idea of a new invention he gives no thought to the limitations of time, or man, or effort. Having his body and mind in complete subjection through iron nerves, he settles down to experiment with ceaseless, tireless, unwavering patience, never swerving to the right or left nor losing sight of his purpose. Years may come and go, but nothing short of success is accepted.

A good example of this can be found in the development of the nickel pocket for the storage battery, an element the size of a short lead-pencil. More than five years were spent in experiments costing upward of a million dollars to perfect it. Day after day was spent on this investigation, tens of thousands of tubes and an endless variety of chemicals were made, but at the end of five years Edison was as much interested in these small tubes as when the work was first begun.

So far as work is concerned, all times are alike to Edison, whether it be day or night. He carries no watch, and, indeed, has but little use for watches or clocks except as they may be useful in connection with an experiment in which time is a factor. The one idea in mind is to go on with the work incessantly, always pushing steadily onward toward the purpose in view, with a relentless disregard of effort or the passage of time.

THOMAS ALVA EDISON—1911

A second and very marked characteristic of Edison's personality is an intense and courageous hopefulness and self-confidence, into which no thought of failure can enter. The doubts and fears of others have absolutely no weight with him. Discouragements and disappointments find no abiding place in his mind. Indeed, he has the happy faculty of beginning the day as open-minded as a child, yesterday's discouragements and disappointment discarded, or, at any rate, remembered only as useful knowledge gained and serving to point out the fact that he had been temporarily following the wrong road.

Difficulties seem to have a fascination for him. To advance along smooth paths, meeting no obstacles or hardships, has no charm for Edison. To wrestle with difficulties, to meet obstructions, to attempt the impossible—these are the things that appear to give him a high form of intellectual pleasure. He meets them with the keen delight of a strong man battling with the waves and opposing them in sheer enjoyment.

Another marked characteristic of Edison is the fact that his happiness is not bound up in the making of money. While he appreciates a good balance at his banker's, the keenness of his pleasure is in overcoming difficulties rather than the mere piling up of a bank account. Had his nature been otherwise, it is doubtful if his life would have been filled with the great achievements that it has been our pleasure to record.

In a life filled with tremendous purpose and brilliant achievement there must be expected more or less of troubles and loss. Edison's life has been no exception, but, with the true philosophy that might be expected of such a nature, he remarked recently: "Spilled milk doesn't interest me. I have spilled lots of it, and, while I have always felt it for a few days, it is quickly forgotten, and I turn again to the future."

Edison to-day has a fine physique, and, being free from serious ailments, enjoys a vigorous old age. His hair has whitened, but it is still abundant, and though he uses glasses for reading, his gray-blue eyes are as keen and bright and deeply lustrous as ever, with the direct, searching look in them that they have ever worn.

Edison in his 'eighties still has a fine physique, weighs over one hundred and sixty-five pounds, and has varied little as to weight in the last forty years. He is very abstemious, hardly ever touching alcohol and caring little for meat. In fact, the chief article of his diet is warm milk, which he finds satisfactory for his need.

He believes that people eat too much, and governs himself accordingly. His meals are simple, small in quantity, and take but little of his time at table. If he finds himself varying in weight he will eat a little more or a little less in order to keep his weight constant.

As to clothes, Edison is simplicity itself. Indeed, it is one of the subjects in which he takes no interest. He says: "I get a suit that fits me, then I compel the tailors to use that as a jig, or pattern, or blueprint, to make others by. For many years a suit was used as a measurement; once or twice they took fresh measurements, but these didn't fit, and they had to go back. I eat to keep my weight constant, hence I never need changed measurements."

This will explain why a certain tailor had made Edison's clothes for twenty years and had never seen him.

In 1873 Mr. Edison was married to Miss Mary Stilwell, who died in 1884, leaving three children—Thomas Alva, William Leslie, and Marion Estelle.

Mr. Edison was married again in 1886 to Miss Mina Miller, daughter of Mr. Lewis Miller, a distinguished pioneer inventor and manufacturer in the field of agricultural machinery, and equally entitled to fame as the father of the "Chautauqua idea," and the founder with Bishop Vincent of the original Chautauqua, which now has so many replicas all over the country. By this marriage there are three children—Charles, Madeline, and Theodore.

For over twenty years Edison's happy and perfect domestic life has been spent at Glenmont, a beautiful property in Llewellyn Park, on the Orange Mountain, New Jersey. Here, amid the comforts of a beautifully appointed home, in which may be seen the many decorations and medals awarded to him, together with the numerous souvenirs sent to him by foreign potentates and others, Edison spends the hours that he is away from the laboratory. They are far from being idle hours, for it is here that he may pursue his reading free from interruption.

His hours of sleep are few, not more than six in the twenty-four, and not as much as that when working nights at the laboratory. In a recent conversation a friend expressed surprise that he could stand the constant strain, to which Edison replied that he stood it easily, because he was interested in everything. He further said: "I don't live with the past; I am living for to-day and to-morrow. I am interested in every department of science, art, and manufacture. I read all the time on astronomy, chemistry, biology, physics, music, metaphysics, mechanics, and other branches—political economy, electricity, and, in fact, all things that are making for progress in the world. I get all the proceedings of the scientific societies, the principal scientific and trade journals, and read them. I also read some theatrical and sporting papers and a lot of similar publications, for I like to know what is going on. In this way I keep up to date, and live in a great, moving world of my own, and, what's more, I enjoy every minute of it."

In conversation Edison is direct, courteous, ready to discuss a topic with anybody worth talking to, and, in spite of his deafness, an excellent listener. No one ever goes away from him in doubt as to what he thinks or means, but, with characteristic modesty, he is ever shy and diffident to a degree if the talk turns on himself rather than on his work.

He is a normal, fun-loving, typical American, ever ready to listen to a new story, with a smile all the while, and a hearty, boyish laugh at the end. He has a keen sense of humor, which manifests itself in witty repartee and in various ways.

In his association with his staff of experimenters the "old man," as he is affectionately called, is considerate and patient, although always insisting on absolute accuracy and exactness in carrying out his ideas. He makes liberal allowance for errors arising through human weakness of one kind or another, but a stupid mistake or an inexcusable oversight on the part of an assistant will call forth a storm of contemptuous expression that is calculated to make the offender feel cheap. The incident, however, is quickly a thing of the past, as a general rule.

If there is anything in heredity, Edison has many years of vigor and activity yet before him. What the future may have in store in the way of further achievement cannot be foreshadowed, for he is still a mighty thinker and a prodigy of industry and hard work.

XXVI

EDISON'S NEW PHONOGRAPH

As related in a preceding chapter of this work, the first commercial phonograph was of the wax cylinder type. Celluloid afterwards superceded wax as a material for the cylinder record, because of its indestructibility. Edison's work on the disc phonograph and record, invented by him in 1878, is related in the following pages.

From the time of his conception of the phonograph in 1877 to the present day Edison has had a deep conviction that people want good music in their homes. That this is not a conviction founded upon commercialism may be appreciated on reading his own words: "Of all the various forms of entertainment in the home, I know of nothing that compares with music. It is safe and sane, appeals to all finer emotions, and tends to bind family influences with a wholesomeness that links old and young together. If you will consider for a moment how universally the old 'heart songs' are loved in the homes, you will realize what a deep hold music has in the affections of the people. It is a safety-valve in the home."

Throughout the years that followed the advent of the earlier type of phonograph with the cylindrical wax records Edison never lost sight of his determination to make it a more perfect instrument, for, of all the children of his brain, the phonograph seems to be the one he loves most. He is the most severe critic of his own work and is never content with less than the best obtainable.

Thus it came about that, some thirteen years ago, having reached the apex of his dissatisfaction with what he thought were the shortcomings of the phonograph and records of that time, he began work on a long-cherished plan of refining the machine and the records so that he could reproduce music, vocal and instrumental, with all its original beauty of tone and sweetness—in fact, a true "re-creation." As the world knows, he has succeeded.

With his characteristic vigor and earnestness Edison plunged into this campaign, fully realizing the immense difficulties of the task he had undertaken. In order to accomplish the desired end he must, in the first place, devise entirely new types of recorder and reproducer which would have essentially different characteristics from any then in existence. In addition to this, an entirely new material must be found and adapted for the surface of the records, a material pliable, indestructible, and, above all, so exceedingly smooth that there should be no rasping, scratching sounds to mar the beauty of the music.

In planning this campaign Edison had decided to return to the disc type of machine and record, which he had invented away back in 1878, and which he now took up again, as it would afford him the greatest scope for his latest efforts.

While simultaneously carrying on a formidable line of experiments to produce the desired material for the records he labored patiently through the days and away into the nights for many months in evolving the new recorder and reproducer, pausing only to snatch a few hours of sleep, which sometimes would be taken at home and at other times on a bench or cot in the laboratory. After some thousands of experiments, extending over a period of more than ten months and conducted with the never-wearying patience so characteristic of him, he perfected his recorder and the diamond-point reproducer which gave him the results for which he strove so many years. This was on the eve of his departure for Europe in August, 1911.

When Edison thinks he has perfected any device his next step is to find out its weakness by trying his best to destroy it. Illustrative of this there may be quoted two instances of severe tests in connection with his alkaline storage battery. After completing it he rigged up a device by means of which a set of batteries were subjected to a series of 1,700,000 severe bumps in the effort to destroy them. When this failed, they were mounted on a heavy electric car, which was propelled with terrific force a number of times against a heavy stone wall, only to show that they were proof against injury by any such means.

His new phonograph reproducer was not exempted from this policy of attempted destruction, and before leaving for Europe he gave instructions for a grilling test, which was, of course, carried out faithfully, but the diamond point was found to be uninjured after playing records more than four thousand times. With such results he deemed it a safe proposition.

On his return from Europe in October, 1911, Edison resumed his attack on the evolution of the new indestructible disc record with a smooth surface, the main principles of which had been determined upon before his departure. In addition, there arose the problem of manufacturing such records in great quantities. The difficulties that confronted him completely baffle description. The whole battle was carried on with the aid of powerful microscopes, which even at their best would fail to reveal the obscure cause of temporary discomfiture. Differences in material, dirt, dust, temperature, water, chemical action, thumb marks, breath marks, cloth and brush marks, and a host of major and minor incidentals, were patiently and painstakingly investigated with a thoroughness that is almost beyond belief to the layman.

Day and night the work was carried on incessantly. During the height of the investigation, toward the close of this five-year campaign, Edison and a few of his faithful experimenters—facetiously called "The Insomnia Squad"—stayed steadily at the works for a period of over five weeks—eating, drinking, working, and sleeping (occasionally) there. During that time Edison went home only four or five times, and then merely to change his clothing. He and the men slept for short periods in the works or in the library, on benches and tables, resuming their labors immediately on waking up. Edison had arranged for an abundant supply of good substantial food which they themselves cooked, hence the inner man was well cared for. The wives of the men came around at intervals with changes of clothing for their husbands. This intense application to work left no time for shaving, with the result that all hands might well have been taken for a gang of traditional pirates from their unkempt appearance.

They were all happy, however, and, strange to say, all increased in weight, although a contrary result might naturally have been expected. The intense work has never ceased, but there has been no similar protracted siege since, as the main principles were practically settled at that time. The foregoing instance has been merely mentioned to illustrate the fierce vigor with which Edison works when he is seeking to complete one of his inventions. He has been, and still is, prosecuting his labors with the same energy to bring about the utmost perfection that is possible.

He has not confined his work to the refinement of the merely mechanical parts, such as the instrument and the records, but during the last ten years he has devoted an immense amount of time to music itself. Becoming convinced that the public desired really beautiful music, he set himself to a thorough study of the subject, not only of compositions, but also of the human voice, its powers and limitations, and of different effects of various styles of orchestration. He determined to hear for himself music of all kinds, and with this object in view hired a number of sight-reading players and singers to render musical selections by the hour.

"THE INSOMNIA SQUAD"—Copyright by Thomas A. Edison

In the past ten years he has heard upward of twenty-five thousand compositions of a wide range, from grand opera to ragtime. As he hears them he indicates his opinions, which range from "beautiful" to "punk," according to his idea of availability for the phonograph. An elaborate card system preserves these indications for further application in selecting music for the phonograph.

It might seem dogmatic to have the reproduction of musical compositions depend upon his opinion, but it must be said that he is not entirely committed to such drastic measures if there is a real demand for some musical selection which does not seem to merit his good opinion. His decision as to a composition is not based on a merely personal whim or fad, but upon his opinion of it from the standpoint of an inventor. He has said to the writer more than once: "There is invention in music just as much as in the arts. Composers such as Verdi, Rossini, Bellini, Donizetti were inventors. They did not copy, nor did some of the other great composers. But the rank and file of musicians are not inventors; they have copied the ideas of the others, consciously or unconsciously. If you will sit down for a few hours and have a lot of miscellaneous compositions played you will be convinced of it."

Edison has had no musical training, as the term is generally understood, and the writer must confess that before hearing the above expression he failed to comprehend the true basis of the inventor's opinions of the various compositions played or sung for him. On several occasions he therefore arranged (unknown to Edison) to have one or more compositions played or sung again after a lapse of some weeks, to see whether or not there would be any similarity of opinion to that first indicated. In every case Edison's judgment was practically, and in some cases precisely, the same as before, thus proving that the opinion first given was not merely a whim, but was based upon some definite line of thought in the inventor's brain.

His excursion into the musical realm has also included the personal hearing of many singers so as to determine their fitness for making phonograph records. This proved to be a wonderfully interesting field of investigation, and he has given a great deal of time to it, listening critically to each voice, good, bad, or indifferent, and patiently writing out his criticism in each case. Not only has he heard a large number of singers who have visited the laboratory for the purpose, but he also had a representative scouring Europe for voices several years ago. This man visited the principal cities and towns of Europe and took phonograph records of the voices of the operatic and other prominent singers in each place and shipped them over to Edison, who listened to each one and recorded his opinion in a series of note-books kept for the purpose. He has in the laboratory at Orange nearly two thousand voice records of this kind. All this is done with the object of securing the really best voices in the world. Probably this is the most unique "voice library" in existence.

He is very deaf, but has a wonderfully acute inner ear, which, being protected by his deafness from the ordinary sounds of life, will catch minute imperfections that are imperceptible to the person of ordinary hearing. In listening to a voice he uses a peculiarly shaped horn which is held close to the ear, and such is the acuteness of his hearing that he at once distinguishes minute changes of register, extra waves, tremolo, non-periodic vibrations, and other minor defects that detract from the true beauty of vocal sounds. In addition, he can immediately recognize the number of overtones and rate of tremolo, which may afterward be verified by a microscopic examination of a record of the same voice.

Edison contends that the phonograph will give the "acid test" of a voice, for it will record nothing more and nothing less than what is in the voice itself, and the record is unchangeable. In his judgment, operatic voices are not necessarily the most perfect ones, for, as he says: "the vocal cords of opera singers are always at the straining-point. They usually sing on roomy stages in large theaters with a large orchestra in front of them, and their voices must go out above all these instruments so as to be heard to the farthest limits of the house. Consequently, they are always doing their utmost and their vocal cords become adapted to heavy work only. People often wonder why their favorite operatic singers do not charm them as much in concert or through the phonograph as they did at the opera, but do not stop to think of the difference between the opera-house and the concert-hall or parlor. I don't mean to say a word of detraction in regard to operatic singers, for I have a great admiration for their wonderful art and for many of their voices, and a great number of them have now recognized the value of special effort to acquire the distinct art and technique of singing for the phonograph (which is a parlor instrument), and have made some really beautiful records."

The writer was one day discussing with Edison the temperament of singers generally and the good opinion that each one usually has of his or her own voice irrespective of any artistic use he or she could make of it. He said: "I don't see what they have to be conceited about. The Almighty has given them a little piece of meat in their throats that differs slightly from the corresponding piece of meat in somebody else's throat. They can take no credit for that, but if they use their brains to interpret and perfect the use of what has been given them, they have accomplished something. What I want is voices that will stand the test of the phonograph and give permanent pleasure to people, irrespective of stage environment, or the press agent, or pleasing personality."

This chapter could be extended to a great length in setting forth the results of Edison's deep study of music which he undertook solely for the purpose of bringing his latest achievement up to the high standard which he set for it so many years ago, but enough has been said to indicate the immense amount of work he has done and the trend of his ideas. That he has been able, amid the round of his multitudinous duties and work, which occupy his time and attention from sixteen to eighteen hours a day, to delve into the subject so profoundly and to evolve ideas that are confessedly awakening the musical world is sufficient to indicate that in spite of his years and herculean labors in the past he has not lost any of the vim or pertinacity that have so distinguished him in days gone by.

XXVII

EDISON'S WORK DURING THE WAR

With the shattering of the world's peace by the great conflict which commenced on July 28, 1914, there came a universal disturbance of industrial conditions. The Edison industries were not exempt.

Edison's activities during the years of the war were of the same intensely vigorous and energetic nature so characteristic of him throughout his busy life. His work during this period is divisible into two distinct sections: first, the working out of processes and the design and construction of nine chemical and two benzol plants to supply chemicals and materials greatly needed by our country; and, second, his war work for the United States government. We will discuss these in the above order.

For many years before the war America had been a large importer of raw materials and manufactured products from England, Germany, and other European countries. Among these may be mentioned potash, dyes, carbolic acid, aniline oil, and other coal-tar products. After hostilities began the activities of the Allied fleets prevented all exportations by Germany and the Central Powers. On the other hand, England and her allies placed embargoes on the exportation from their countries of all materials and products which could be used for food or munitions of war.

Thus there suddenly came a great embarrassment to numerous American industries. By reason of our continued importation for many years our country had become dependent upon Europe for supplies of various products and had made practically no provision for the manufacture of these products within our own borders.

Inasmuch as our narrative concerns Edison and his work, we shall not attempt to name all the industries thus affected, but will confine ourselves to a mention of the items relating to his own needs and of those which he promptly took steps to produce for the relief of many industries and for the general good of the country. These items were carbolic acid, aniline oil, myrbane, aniline salts, acetanilid, para-nitro-acetanilid, paraphenylenediamine, para-amidophenol, benzidine, benzol, toluol, xylol, solvent naphtha, and naphthaline flakes.

Edison's principal requirements were potash for his storage battery and carbolic acid and paraphenylenediamine for use in the manufacture of disc phonograph records. After a great deal of experimenting he found that caustic soda could be used in his storage battery and therefore employed it until new supplies of potash were obtainable.

Carbolic acid and paraphenylenediamine had been previously imported from England and Germany and as there was practically none produced in the United States and no possibility of substituting other products Edison realized that he would be compelled to manufacture them himself, as the source of supply was cut off. He, therefore, as usual, gathered together all available literature and plunged into a study of manufacturing processes and quickly set his chemists to work on various lines of experiment.

Having decided through these experiments on the process by which he would manufacture carbolic acid synthetically, Edison designed his first plant, gathered the building material and apparatus together and instructed his engineers to rush the construction as fast as possible. By working gangs of men twenty-four hours a day the plant was rapidly completed and on the eighteenth day after the work of construction was begun it commenced turning out carbolic acid. Within a month this plant was making more than a ton a day and gradually increased its capacity until, a few months afterward, it reached its maximum of six tons a day.

It soon became publicly known that Edison was manufacturing carbolic acid, and he was overwhelmed with offers to purchase the excess over his own requirements. The demand for carbolic acid became so great that he decided to erect a second plant. This was quickly constructed and its capacity, which was also six tons per day, was contracted for before the plant was fully completed. It is interesting to note that the army and navy departments of the United States were among the first to make long contracts with Edison for his carbolic acid, from which they made explosives that were badly needed.

We must digress here to show an emergency that had arisen during the early days of the first carbolicacid plant. There had come about a serious shortage of benzol, which is a basic material in the manufacture of synthetic carbolic acid. Benzol is a product derived from the gases arising from the destructive distillation of coal in coke ovens. At the time of which we are writing (beginning of 1915) there was only a comparatively small quantity of benzol produced in the United States.

Mr. Edison realized that without a continuous and liberal supply of benzol he would be unable to carry out his project of producing carbolic acid in large quantities. He had also been approached by various textile manufacturers to make aniline oil, which was essential to their continuance in business, and of which there was practically no supply in the country. Without it he could not make paraphenylenediamine. Benzol is also a basic material in making aniline oil.

Therefore, it became doubly important to arrange for an adequate and continuous supply of benzol. Edison made a study of the methods and processes of producing benzol and then made proposals to various steel companies to the effect that he would, with their permission, erect a benzol plant at their coke ovens, operate the same at his own expense, and pay them a royalty for every gallon of benzol, toluol, xylol, or solvent naphtha taken from their gases. Such arrangement would not only meet his requirements, but at the same time would give the steel companies an income from something which they had been allowing to pass away into the air. He succeeded in making arrangements with two of the companies—namely, the Cambria Steel Company at Johnstown, Pennsylvania, and the Woodward Iron Company, Woodward, Alabama.

Ordinarily, it requires from nine to ten months to erect a benzol plant, but before making his proposal to the steel companies Edison had worked out a plan for erecting a practical plant within sixty days, and had laid it out on paper. He was sure of his grounds, because from his vast experience he knew where to pick up the different pieces of apparatus in various parts of the country.

The contract for his first benzol plant at Johnstown, Pennsylvania, was signed on January 18, 1915, and the actual work was begun an hour after the contract was signed, with the final result that in forty-five days afterward the benzol plant was completed and commenced working successfully. The second plant, at Woodward, Alabama, was completed within sixty days after breaking ground, the two weeks difference in time being accounted for by the fact that Woodward was farther away from the base of supplies and there were delays in railroad transportation of materials.

Being sure, through these contracts, of a continuous supply of benzol, Edison designed a plant for making aniline oil. By working gangs of men day and night, the erection of this plant was completed in forty-five days. The capacity of the plant, four thousand pounds per day, was fully contracted for by anxious manufacturers long before the machinery was in place.

Let us now consider Edison's work on paraphenylenediamine. This is a chemical product which is largely used in dyeing furs black. America had imported all her requirements from Germany, but within a few months after the beginning of hostilities the visible supply was exhausted and no more could be expected during war-times. Fur-dyers were in despair. This product being also absolutely essential in the manufacture of phonograph records, Edison worked out a process for making it, and as his requirements were very moderate he established a small manufacturing plant at the Orange laboratory and soon began to produce about twenty-five pounds a day. In some way the news reached the ears of many desperate fur-dyers, and Edison was quickly besieged with most urgent requests for such portion of his output as could be spared. Fortunately, a small proportion of the output was available and was distributed daily in accordance with the necessities of those concerned. This small quantity being merely a drop in the bucket, the fur-dyers earnestly besought Edison to establish a larger plant and supply them with greater quantities of paraphenylenediamine, as their business had come almost to a standstill for lack of it. He, therefore, designed and constructed rapidly a larger plant, which, when put into operation, was soon producing two hundred to three hundred pounds a day, thus saving the situation for the fur-dyers. The capacity of this plant was gradualy increased until it turned out upward of a thousand pounds a day, of which a goodly proportion was exported to Europe and Japan.

Lack of space has prevented the narration of more than a mere general outline of some of Edison's important achievements during part of the war years along chemical and engineering lines and in furnishing many of the industries of the country with greatly needed products that, for a time at least, were otherwise unobtainable. Much could be written about his work on producing myrbane, aniline salts, acetanilid, para-nitro-acetanilid, para-amido-phenol, benzidine, toluol, xylol, solvent naphtha, and naphthaline flakes—how his investigations and experiments on them ran along with the others, team fashion, so to speak, how he brought the same resourcefulness and energy to bear on many problems, and how he eventually surmounted numerous difficulties—but limitations of space forbid. Nor can we make more than a mere passing mention of the assistance he gave to the governments in the quick production of toluol and in furnishing plans and help to construct and operate two toluol plants in Canada. Suffice it to say that his achievements during this episode in his career were fully in accord with the notable successes he had already scored. It may be noted that in the three years following 1914 others went into the business of manufacturing the above chemicals, and as they installed and operated plants and furnished supplies needed in the industries Edison withdrew and shut down his special plants one after another.

Let us now take a brief glance at the patriot-inventor at work for his government in war-times and especially during the last two years of the Great War.

In the late summer of 1915 the Secretary of the Navy, Hon. Josephus Daniels, communicated to Mr. Edison an idea he had conceived of gathering together a body of men preeminent in inventive research to form an advisory board which should come to the aid of our country in an inventive and advisory capacity in relation to war measures. In this communication Secretary Daniels made an appeal to Edison's patriotism and asked him to devote some of his effort in the service of the country and also to act as chairman of the board. Although he was already working about eighteen hours a day, Edison signified his consent. In the fall of 1915 the board was organized and subsequently became known as the Naval Consulting Board of the United States. Mr. Edison was at first chairman and subsequently became president of the board.

The history of the work and activities of the board is too extensive to be related here in detail and can only be hinted at. Indeed, it is the subject of a separate volume which is being published by the Navy Department. We shall, therefore, confine our narrative to the story of Edison's work.

In December, 1916, Secretary Daniels expressed a desire that Mr. Edison visit him in Washington for an important conference. At that time it seemed almost inevitable that the United States would be drawn into the conflict with Germany sooner or later, and at the conference Secretary Daniels asked Edison to devote more of his time to the country by undertaking experiments on a series of problems, a list of which was handed to him.

Edison signified his assent, agreeing to give his whole time to the government without charge, and returned to his laboratory. He immediately put everything else aside, and with characteristic enthusiasm and energy delved into the work he had undertaken. The problems referred to covered a wide range of the sciences and arts, and time being an essential element, he added to his laboratory staff by gathering together from various sources a number of young men, experts in various lines, to assist him in his investigations.

Inasmuch as Edison's war work for the government occupied his entire time for upward of two years, it is manifestly out of the question to narrate the details within the limits of a chapter. We must, therefore, be content to itemize the principal problems upon which he occupied himself and assistants and as to which he reported definite results to Washington. The items are as follows:

1. Locating position of guns by sound-ranging.

2. Detecting submarines by sound from moving vessels.

3. Detecting on moving vessels the discharge of torpedoes by submarines.

4. Quick turning of ships.

5. Strategic plans for saving cargo boats from submarines.

6. Collision mats.

7. Taking merchant-ships out of mined harbors.

8. Oleum cloud shells.

9. Camouflaging ships and burning anthracite.

10. More power for torpedoes.

11. Coast patrol by submarine buoys.

12. Destroying periscopes with machine-guns.

13. Cartridge for taking soundings.

14. Sailing-lights for convoys.

15. Smudging sky-line.

16. Obstructing torpedoes with nets.

17. Under-water search-light.

18. High-speed signaling with search-lights.

19. Water-penetrating projectile.

20. Airplane detection.

21. Observing periscopes in silhouette.

22. Steamship decoys.

23. Zigzagging.

24. Reducing rolling of warships.

25. Obtaining nitrogen from the air.

26. Stability of submerged submarines.

27. Hydrogen detector for submarines.

28. Induction balance for submarine detection.

29. Turbine head for projectile.

30. Protecting observers from smoke-stack gas.

31. Mining Zeebrugge harbor.

32. Blinding submarines and periscopes.

33. Mirror-reflection system for warships.

34. Device for look-out men.

35. Extinguishing fires in coal bunkers.

36. Telephone system on ships.

37. Extension ladder for spotting-top.

38. Preserving submarine and other guns from rust.

39. Freeing range-finder from spray.

40. Smudging periscopes.

41. Night glass.

42. Re-acting shell.

It will be seen that Mr. Edison's inventive imagination was permitted a wide scope. He fairly reveled in the opportunity of attacking so many difficult problems and worked through the days and nights writh unflagging enthusiasm. He committed his business interests to the care of his associates, and during the two years of his work for the government kept in touch with his great business interests only by means of reports which were condensed to the utmost. In addition, for two successive winters, he gave up his regular winter vacation on his Florida estate, usually a source of great enjoyment to him. But it was all done willingly and without a word of regret or dissatisfaction so far as the writer's knowledge goes.

Although we cannot take space to discuss the above items in detail, the reader will probably have a desire to know something of Edison's work in regard to the submarines.

In view of the vast destruction of shipping, perhaps it is not an overstatement to say that the most vital problem of the late war was to overcome the menace of the submarine.

Undoubtedly there was more universal study and experiment on means and devices for locating and destroying submarines than on any other single problem.

The class of apparatus most favored by investigators comprised various forms of listening devices by means of which it was hoped to detect and locate by sound the movement of an entirely submerged submarine. The difficulties in obtaining accurate results were very great even when the observing vessel was motionless, but were enormously enhanced on using listening devices on a vessel under way, on account of the noises of the vessel itself, the rushing of the water, and so on.

Edison's earliest efforts were confined to the induction balance, but after two months of intensive experimenting on that line he gave it up and entered upon a long series of experiments with listening devices, employing telephones, audions, towing devices, resonators, etc. The Secretary of the Navy provided Edison with a 200-foot vessel for his experiments, and in the summer and fall of 1917 they had progressed sufficiently to enable him to detect sounds of moving vessels as far distant as five thousand yards. This, however, was when the observing vessel was at anchor. The results with the vessel under way, at full speed, were not poor.

Having pushed the possibilities along this line to their reasonable limit, Edison was of the opinion that this plan would not be practical and he turned his thoughts to another solution of the problem—namely, to circumvent the destructive operation of the submarine and avoid the loss of ships. He had discovered in his experimenting that the noise made by a torpedo in its swift passage through the water was very marked and easily distinguishable from any other sound.

With this fact as a basis, Edison, therefore, evolved a new plan, which had two parts: first, to provide merchant-ships with a listening apparatus that would enable them, while going at full speed, to hear the sound of a torpedo as soon as it was launched from a submarine; and, second, to provide the merchant-ships with means for quickly changing their course to another course at right angles. Thus, the torpedo would miss its mark and the merchantship would be saved. If another torpedo should be launched, the same tactics could be repeated.

His further investigations were conducted along this line. After much experimenting he developed a listening device in the form of an outrigger suspended from the bowsprit. This device was so arranged that it hung partly in the water and would always be from 10 to 20 feet ahead of the vessel, but could be swung inboard at any time. The device was about 20 feet long and about 16 inches in width and was made of brass and rubber. It contained brass tubes, with a phonograph diaphragm at the end which hung in the water. The listening apparatus was placed in a small room in the bow of the vessel. There were no batteries used. With this listening apparatus, and while the vessel was going full speed, moving boats 1,000 yards away could be easily heard in rough seas. This meant that torpedoes could be heard 3,000 yards away, as they are by far the noisiest craft that "sail" the ocean.

The second step in Edison's plan—namely, the quick changing of a ship's course, was accomplished with the "sea anchor." This device consists of a strong canvas bag which is attached to a ship by long ropes. When thrown overboard the bag opens, fills with water, and acts as a drag on a ship under way. Edison's plan was to use four or more sea anchors simultaneously. In a trial made with a steamship 325 feet long, draught 19 feet 6 inches, laden with 4,200 tons of coal, by the use of four sea anchors, the vessel going at full speed, was turned at right angles to her previous course with an advance of only 200 feet, or less than her own length. This means that if an enemy submarine had launched a torpedo against the ship while she was on her original course it would have passed by without harming her, thus making submarine torpedo attack of no avail. It may be noted parenthetically that this apparatus has its uses in the merchant-marine in peacetimes. For instance, should the look-out on a steamship running at full speed sight an iceberg 300 or 400 feet ahead this device could be instantly put into use and the ship could be turned quickly enough to avoid a collision.

EDISON AT WORK ON RUBBER EXPERIMENTS.
FROM A MOVING PICTURE TAKEN DECEMBER,
1928

There is only space for a passing mention of the immense amount of data which Edison gathered, tabulated, and charted in his study and evolution of strategical plans suggested by him to the government in the line of lessening the destruction by submarines. He spent day and night for several months with a number of assistants working out these plans. It is not possible to make more specific mention of them here, as they are too voluminous for these pages.

With this tremendous amount of work pressing on him he retained his accustomed good health and buoyancy, due, undoubtedly, to his cheerful spirit, philosophical nature, and abstemious living. Soon after the armistice was signed his experimental work for the government came to an end, and he then switched back to the general supervision of his business interests and to his ceaseless experiments through which he is continually making improvements and refinements in the products of the large industries which he established and in which he is so greatly interested.

Mention should also be made of another extensive project he has undertaken, and that is the production of rubber from plants, weeds, bushes, shrubs, etc., grown in the United States. This he speaks of as "emergency" rubber, to be resorted to in case our country should ever be embarrassed in obtaining a supply of rubber from present sources. This is a tremendous problem, but he is applying to its solution the same resourceful powers that have characterized his previous endeavors.

Herein, and in the development of new ideas, lies Edison's daily work and pleasure, and although he is in his eighties at this writing, with still boundless energy, it may be said of him

"Age cannot wither
him, nor custom
stale
His infinite
variety."